THE DEVELOPMENT OF PHYSICAL THEORY
IN THE MIDDLE AGES

The Development of
Physical Theory
in the Middle Ages

James A. Weisheipl

Ann Arbor Paperbacks
THE UNIVERSITY OF MICHIGAN PRESS

CONTENTS

INTRODUCTION

One radiant morning in the spring of 1591, so the story goes, a young, impertinent professor of mathematics at the University of Pisa assembled a throng of professors, undergraduates and townsmen in the cathedral square to witness an experiment. He was determined to put an end to a controversy which had long raged between himself and the older professors. The aged savants of philosophy scoffed; undergraduates were hilarious, and the townsmen curious about the whole demonstration. Before their eyes the young Galileo climbed the spiral staircase of the near-by Leaning Tower, carrying with him two balls of lead, one a hundred times heavier than the other. When he had reached the uppermost gallery, he carefully balanced the two balls on the parapet and rolled both over the edge simultaneously. To the astonishment of all assembled the two weights reached the ground with a single thud. In that fall Aristotle and the aged Aristotelians were henceforth discredited and the birth of modern science was heralded throughout Europe.

According to another story the same professor, who by this time had accepted a professorship at the famous University of Padua, contrived to make a simple eyeglass, or telescope, whereby he could study the movement of the moon, planets and stars. Turning his home-made telescope towards the sun, he excitedly noticed that this celestial body was covered with dark spots which gradually narrowed as they approached the edge of the disk. He hurriedly invited an Aristotelian professor of philosophy to see for himself that the celestial bodies are not perfect as Aristotle had claimed and that, in fact, the Copernican theory of the earth's motion around the sun is clearly proved by sight alone. It is said that Galileo's Aristotelian associate refused to gaze through the telescope, insisting that the senses cannot contradict what Aristotle had demonstrated by reason. Nevertheless the downfall of Aristotle had at long last been secured for the modern world, even though Galileo himself was to be persecuted by unscrupulous churchmen for his "crime" against the established philosophy of the schools and the infallibility of dogma.

Whatever may be said of these familiar popular pictures, the belief that Galileo Galilei, almost singlehanded, procured the "downfall of Aristotle" and with paternal solicitude ushered modern science into a world engulfed in the "Aristotelian morass" (to use a recent description) cannot be entirely dismissed as groundless fiction. Galileo himself is partly responsible for the impression that he is the originator of "an entirely new science in which no one else, ancient or modern, has discovered any of the most remarkable laws which I demonstrate to exist in both natural and violent movement."[1]   His habitual braggadocio and not infrequent attacks upon the physics of Aristotle suggest that he recognized no debt to his predecessors, or at least that he would not willingly do so in public. There can be no doubt, too, that the Galileo legend has been made much of by those who interpret his achievements as a victory over authority in general and the Roman Church in particular. Discussions concerning papal infallibility and intellectual freedom in scientific investigation still bog down over the "crime of Galileo". But it would seem that the most common source of the legend of singlehanded innovation, or at least of the perpetuation of it even in the most recent textbooks of science, is an almost universal

---

[1] Letter of Galileo to Belisario Vinta (1610), trans. by S. Drake, *Discoveries and Opinions of Galileo*, 1957, p. 63.

lack of knowledge of the nature of scientific theory before Galileo—the "Aristotelian morass" from which he raised himself. The originality of Galileo cannot, of course, be denied. Nor would one wish to deny it. However, this originality cannot be appreciated without understanding the precise point of contrast in physical theory; nor can it be evaluated without knowing the vast fund of scientific knowledge which Galileo inherited from his predecessors. Every work of genius has an ancestry, legitimate or otherwise. Modern readers could learn a great deal from the ancestors of Galileo, particularly from the philosophers and scientists of the Latin Middle Ages who synthesized and elaborated the learning of Antiquity.

For the specialist in the history of medieval science much has been written during the past fifty years on "the precursors of Galileo". These erudite studies for the most part attempt to show how a particular concept of modern classical physics was already anticipated by certain schoolmen of the fourteenth century, or at least foreshadowed by them. Such studies are essential for a proper perspective of scientific ideas. For the general reader, however, it is more useful to grasp the aim and extent of scientific theory as understood by the Christian schoolmen of the Middle Ages. There is much in the medieval view of physical science which merits the attention of modern readers who are prepared to reconsider the perennial problems of the physical universe.

Actually there was no one physical theory universally and indisputably accepted by all scholars in the Middle Ages. Even after the introduction of Aristotle into the Latin West, not all teachers interpreted him in the same way; nor did all unquestioningly accept what they thought the "Philosopher" (Aristotle) had said. Certain views of Aristotle, such as the eternity of the universe, the indifference of divine Providence toward human affairs, and various minor statements, were opposed to the Christian faith, and hence were generally rejected. Then, too, there were many non-Aristotelian influences which consciously or unconsciously moulded individual views. The impassioned controversies of the Middle Ages concerning physical theory were not academic exercises conducted "within the framework of the Christian faith" and "regulated by Church authority"; they were radically conflicting views, having nothing to do with dogma, proposed by Christian philosophers, each of whom was convinced he had the true explanation of physical reality. It would be useless, if not impossible, for the general reader to examine all the scientific works written in the Middle Ages for a general physical theory; and it would be a monstrous distortion to select the least common denominator unanimously accepted by all schoolmen. However, if we direct our attention principally to Albertus Magnus, the most experienced scientist, and to St. Thomas Aquinas, the "Common Doctor", we shall at least be considering the greatest scientific minds produced by the Middle Ages. Our purpose here is to sketch the general theory of physical science as understood by these two "preeminent men in philosophy."[2] It will be necessary, however, to trace its origins in the Latin West and to follow its fortunes to the time of Galileo.

EARLY MEDIEVAL SCIENCE

The first four centuries of the Christian era were predominantly Greek in language, thought and culture. The original distrust of Greek philosophy and science displayed by the Early Christians was to a certain degree displaced by the more intellectual synthesis introduced by converts from educated Greek paganism. For thinkers such as St. Justin Martyr, St. Clement of

[2] Siger of Brabant, *De Anima Intellectiva*, iii, ed. P. Mandonnet, *Siger de Brabant*, vol. ii, Louvain, 1908, p. 152.

Alexandria and Origen, the "folly of the Cross" was the true philosophy, the true *gnosis,* and whatever happened to be good and true in Greek philosophy was derived from Moses, and therefore rightly belonged to the Christians. The philosophical thought of the Mediterranian world during the second and third centuries was predominantly Platonic in metaphysics, Aristotelian in logic and Stoic in morality. Such technical sciences and arts as were known at the time were the common property of Christians and pagans, Greeks and Romans. But there was hardly any serious effort to develop a physical theory of nature. The scientific labours of Aristotle were to a large extent abandoned by his disciples, who preferred grammar, logic and metaphysics to the detailed study of nature. One notable exception, however, was the study of medicine, which was greatly enhanced by the vast labours and prolific writings of Galen of Pergamum in the second century after Christ.

It was not so much natural studies as mathematics which interested the Mediterranean world at this time. The universally respected Euclidean geometry had been applied successfully to diverse problems of mechanics and statics by Archimedes of Syracuse (*c.* 287-212 B.C.). The application of mathematics to physical problems reached new heights during the second century of the Christian era, largely through the efforts of Hero of Alexandria in mechanics and in optics, and through the genius of Claudius Ptolemy in astronomy. Perhaps it was due to the success of these mathematical sciences, namely astronomy, optics and mechanics, that the Platonic and Pythagorean view of nature was preferred to the more naturalistic and empirical view of Aristotle. C. Singer and others have somewhat inaccurately called this mathematical approach to nature the "Alexandrian divorce of science and philosophy". More properly one might say that the Alexandrian scientists rejected the Aristotelian philosophy of nature in favour of the

Platonic. In any case, it was the Platonic view of the universe which was generally preferred, and it was this view which the early Greek Fathers of the Church utilized.

Unlike his predecessors, Plato had been deeply impressed with the stable elements in nature. Individual rabbits, to use a modern example, are begotten and die, but a rabbit would always be an unmistakable species of the hare family, even if there were no rabbits alive. A fossil of a dinosaur is immediately recognized as such, although there is no individual dinosaur existing today. It is the idea, or essential structure, of physical things which is stable and eternal, not the individual manifestations, which are mere passing shadows of the absolutely real. But matter, too, is stable and eternal, for it is "the mother of all becoming"; matter itself never comes into being or ceases to be, but it is the source of all change. Hence for Plato there are only two kinds of absolutely real being: the autonomous,. immaterial ideas and the formless matter out of which all shadows are made. The eternal ideas cast images on a formless, undifferentiated screen like a movie projector; the projected image is only a shadow made up of form (*morphé*), an individual copy of the eternal idea, and matter (*hylé*), the physical stuff out of which the shadow is made. Plato compared matter to a mother, the eternal idea to a father, and "the nature that arises between them to their offspring". (*Timaeus, 50 D.*) He admitted that it is difficult to conceive this unformed matter, which is the receptacle of the projected images. It seems to be at one and the same time like clay out of which statues are made and like the wall of a cave upon which shadows are cast by fire.

> [Matter] must be called always the same; for it never departs at all from its own character; since it is always receiving all things, and never in any way whatsoever takes on any character that is like any

of the things that enter it; by nature it is there as a matrix for everything, changed and diversified by the things that enter it, and on their account it appears to have different qualities at different times, while the things that pass in and out are to be called copies of the eternal things, impressions taken from them in a strange manner that is hard to express. (*Timaeus,* 50 B-C.)

Three significant points should be noticed about Plato's view of the physical universe, the world of shadows. First, matter and form are two distinct *things* making up the transistory phenomenon. Plato could never conceive of matter and form as constituting a uniquely single thing. This was to have important consequences in psychology, where a human being was considered a spirit, or soul, imprisoned in a body. Plato's dichotomy was to haunt not only Christian mysticism, but much of modern psychology since the time of Descartes. Second, in Plato's description matter *out of which* the shadows are made was confused with dimensional space *into which* bodies can be placed. While the early Christian thinkers took no exception to this, Aristotle and medieval philosophers were insistent in pointing out the confusion. The third and most significant scientific point to notice is that Plato failed to account for the *change* in the shadows. If the ideas are eternal and immutable and the screen likewise eternal and immutable, the same image should be projected forever. Plato himself perceived the difficulty with some embarrassment. One solution proposed was to compare the screen to a winnowing basket, which, on being shaken by quick jerks, allows the chaff to settle in one part of the basket and the grain in another; thus, with the shaking of the universal receptacle, similar bodies would tend to congregate. Plato's winnower in the case of the universe was called the "world soul". But in Plato's description the world soul is the final product in the evolutionary process, and, as Aristotle later pointed out, what is posterior cannot be the original cause of change.[3] Plato was left with the unsatisfactory view that the primitive elements of earth, water, air and fire originally moved in all directions by chance.

However, for Plato wisdom could not be found in the study of transistory shadows, but only in the contemplation of the pure ideas. He was particularly impressed by the perfect organization of each autonomous idea, an organization neatly expressed through its logical definition. This organization, revealed in such ideas as that of justice, goodness, man and the individual elements, was derived from its numerical and geometrical structure. Each eternal idea reflected in the physical world was derived from and produced by its geometrical structure, and this in turn was derived from its numerical structure, the ultimate source of all wisdom. Thus the idea of goodness, existing in this universe only in an imperfect and transitory state, but existing in all its purity apart from matter, springs from its geometrical indivisibility, and this in turn is derived from unity, the principle of all number. Plato conceived fire (tetrahedron), air (octahedron) and water (icosahedron) as multiples of the isosceles triangle and he saw in this an explanation of why these elements can so easily be generated one from the other; but earth, being a cube and a multiple of the equilateral triangle, can only be scattered throughout the other elements until like parts meet together and solidify.[4] The conviction that mathematics lay behind all physical phenomena, a conviction which was also that of Pythagoras, led Plato and his disciples to embark on a search for the mathematical structure of the physical world. The study of mathematics, therefore, was for them the key which opened the door of nature and of wisdom. We are told by a

[3] Aristotle, *Metaph.*, xii, c. 6, 1071b 37-1072a 3.
[4] Plato, *Timaeus*, 56 D.

late source[5] that Plato inscribed over the entrance to his school, the Academy, "Let no one ignorant of geometry enter my door."

Although the Babylonians and Egyptians collected astronomical data and developed mathematics to a certain degree of perfection, theoretical astronomy, which is the application of geometrical figures to celestial motions, did not begin until Plato posed the problem to his disciple, Eudoxus.[6] The Platonists, of course, did not think of astronomy as the *application* of mathematics to celestial motion, but as the discovery of the *ultimate cause* of celestial motion. Contemplation of the ultimate causes of the purest ideas was for Plato and his followers the goal of all science, for this contemplation was considered true wisdom.

The early Fathers of the Church had little direct interest in science or in physical theory. However, those who knew Greek were indeed familiar with the prevailing Platonic view, which was given new impetus by Ammonius Saccas, an apostate Christian, and Plotinus, his disciple. Christian scholars, such as Clement of Alexandria, Origen and the later Cappadocian Fathers, found the spiritual beauty of Platonism, even in science, easily compatible with Christianity. Plato's subsistent ideas were interpreted as eternal exemplars in the mind of the Creator; the coeternal spatial screen was thought to be the created chaos from which God created all things physical; and the goal of man was to purify himself of the material things of this world by mortification in order to be united with Christ for all eternity in heaven. The meagre scientific notions expressed by Plato in the *Timaeus* were eagerly utilized by the Cappadocian Fathers in their homilies on the six days of creation.

[5] John Tzetzes, an early twelfth-century Byzantine poet, in his *Chiliad*, 8, 972, quoted by R. E. Moritz, *On Mathematics*, New York, 1942, p. 292.

[6] Simplicius, *Comm. in De Caelo*, quoted by T. L. Heath, *Greek Astronomy*, London, 1932, p. 67.

These cosmographical verses in Genesis (i. 1-26) were commonly known as the *Hexaemeron*, the six days.

Among the Latin Fathers, Hilary of Poitiers, Ambrose and Jerome knew Greek well and they followed the Cappadocians in the moderate use of Platonic physics in explaining the six days of creation. Although St. Augustine knew practically no Greek, he was able to assimilate and adopt significant elements of neo-Platonism into the Christian view of the universe with great genius and dexterity. His literal commentary on Genesis and the last three books of his *Confessions* are a marvellous example of such an assimilation of Plato's cosmology. St. Augustine, of course, was an exceptional genius; even though he borrowed from neo-Platonic sources whatever he thought was true, his discussions concerning time and eternity, change, the nature of first matter and the human soul, are profoundly original.

Pure Greek science, however, would have been completely lost to the early Middle Ages, if it had not been for Boethius. Most of it was indeed lost to the West in the final collapse of the Roman Empire and in the almost universal disruption of learning in the dark centuries which followed. Boethius, a Roman consul and minister of affairs under Theodoric, king of the Ostrogoths, had received an excellent education in Athens. He was familiar with the full heritage of Greek science and philosophy. Aware of the general ignorance of the Latins of his day (*c.* 475-524), he set himself the task of translating all the works of Plato, Aristotle, Euclid, Ptolemy and Archimedes.[7] If he had succeeded in this ambitious plan, the history of science would undoubtedly have been very different. But before he was fifty years of age he was imprisoned near Pavia, tortured and martyred by the Arian king he once served. However, he bequeathed to the early Middle Ages an almost

---

[7] Boethius, *In Lib. Arist. Periherm. Comm.*, ed. 2a, lib. ii, c. 3, and letter of Cassiodorus to Boethius, *Lib. Var.*, i, ep. xv.

complete logic, including translations of at least two of Aristotle's works, a summary of Euclid's *Elements,* together with an elementary handbook of arithmetic and another on music, both drawn from Nicomachus of Gerasa. For more than five hundred years this meagre inheritance served as the basis for the study of the liberal arts. This, together with a partial translation of Plato's *Timaeus* and the commentaries of Chalcidius and Cicero, represented the extent of Greek science available to the Latin West until the renaissance of the twelfth century.

One should not infer from this that there were no schools or scholars in Latin Christendom from the death of Boethius until the mid-twelfth century. There were in fact numerous cathedral, monastic and court schools throughout Europe. And there were great scholars, even a few who could read Greek, as could the Celtic monks and John Scotus Erigena. But their interests were primarily centred on theology, grammar and logic. Men like Gerbert of Rheims, who is said to have acquired some Arabic science in Spain, were extremely rare, and they contributed nothing to a physical theory of the universe or to particular problems of science. Not even the famous cathedral school of Chartres had anything to offer toward a physical theory. Theologians commenting on the Hexaemeron continued to utilize the elementary Platonic cosmology transmitted by Augustine, Ambrose and other patristic writers. It was not until the recovery of Aristotelian science and scientific methodology in the twelfth century that any serious attempt could be made to elaborate a physical theory of nature.

RECOVERY OF GREEK SCIENCE

There were two main channels through which Greek science entered the Latin West about the middle of the twelfth century. The more significant, perhaps, was through translations from the Arabic, coming mainly from Spain, where large territories were reconquered from the Moslems. The more accurate channel, however, was through translations directly from the Greek, coming largely through Italy.

With the conquest of Alexandria and Syria the Mohammedans had come into contact with Greek science, which included the scientific views of Plato and Aristotle. This science, translated into Arabic from Syriac and Greek, was assimilated to a large extent by the Moslem philosophers (Mutazilites, not to be confused in any way with the theologians, or Mutakallûn). But, apart from the exceptionally Aristotelian Averroes, Arab philosophers were generally oblivious to the radical opposition between Plato and Aristotle; the extraordinary synthesis which Arab philosophers produced reflected the neo-Platonism of Plotinus and Proclus more than it did the scientific theory of Aristotle. However, the Arabs made significant contributions to the study of medicine, astronomy and algebra.

After the city of Toledo was reconquered by the Christians in 1085, a wealth of Arabic scientific literature became available to those who could read it. Under the direction of John, Archbishop of Toledo (1151-66), one notable team of translators, Dominic Gundissalinus and Ibn Daud, rendered important works of Avicenna, Alfarabi, Alkindi and Ibn Gabirol into Latin through the medium of the Spanish vernacular. These translations stimulated scholars from all over Europe to hasten to Spain in search of scientific knowledge. The German, Hermann of Carinthia, translated the *Planisphere* of Ptolemy, various astronomical treatises and the fatalistic *Great Introduction to Astrology* by Albumazar, which was to influence Western thought for many centuries. An Englishman, Adelard of Bath, made the first complete translation of the *Elements* of

Euclid, including the spurious Arabic Books XIV and XV. The pseudo-Aristotelian work *On Plants* and *On Minerals* were translated by Alfred of Sareshel, popularly known as "Alfred the Englishman". Other Englishmen joined in the literary boom beyond the Pyrenees, among them Daniel of Morley and Robert of Chester, both of whom wrote popular works incorporating the new scientific ideas.

The most prolific of the twelfth-century translators was an Italian, Gerard of Cremona, who determined to learn Arabic when he chanced upon a reference to the *Almagest* of Ptolemy, which he could not find in Latin. In the ancient catalogues he is credited with having translated seventy-one works from the Arabic. This septuagenarian did not live to see public acclaim for his lifelong efforts. He deserves recognition, however, as the midwife of Western science. To Western medicine he gave Avicenna's *Canons of Medicine* and many works of Galen and Hippocrates. To astronomy he gave Ptolemy's *Almagest* and numerous detailed observations made by the Arabs. To geometry he gave the best complete translation from the Arabic of Euclid's *Elements*. To the philosophy of nature he gave Aristotle's *Physics, De Caelo, De Generatione* and *Meteora* (I-III), including a number of Greek commentaries on Aristotle. To scientific methodology he gave Aristotle's *Posterior Analytics*. One might expect Greek thought which had passed through Syriac and Arabic before donning Latin garb to appear a little fatigued. The fact that it did not suffer complete prostration attests to the solicitude of the numerous translators and copyists.

About the same time, however, more accurate translations of scientific and mathematical works were being made directly from the Greek. These were produced largely in Sicily and in northern Italy. In the Norman kingdom of Sicily, Greek, Latin and Arabic culture mingled freely under Christian rule. Commerce with the

East continued, libraries were enlarged and scientific studies were encouraged personally by the Sicilian kings from the middle of the twelfth century until the time of Frederick II and Manfred. In such a learned and cosmopolitan atmosphere the production of translations was almost inevitable. Henricus Aristippus and Eugene of Palermo, both members of the royal administration, were only two of the many scholars who rendered much of Greek philosophy and mathematics into Latin. Aristippus was the first to translate the *Meno* and *Phaedo* of Plato and the fourth book of Aristotle's *Meteora*. This latter work found immediate circulation in the north, but the two dialogues of Plato apparently passed unnoticed until discovered by Albertus Magnus. From the hands of Eugene came the *Optics* of Ptolemy (from the Arabic), which might otherwise have been completely lost. There can be no doubt that scholars at the Sicilian court were familiar with geometrical analysis and applied mathematics, as found in the most advanced works of Euclid, Ptolemy, Proclus and Hero of Alexandria, long before the northern countries had had time to assimilate the *Elements* of Euclid. But their work contributed more to particular problems of mathematics and astronomy than to a general physical theory of nature.

The most important of the northern translators was James of Venice, who is credited with the earliest Latin version of Aristotle's *Physics, De Anima, Metaphysics* and some shorter physical treatises; he also translated Aristotle's *Prior* and *Posterior Analytics,* the *Topics* and the *Sophistical Refutations,* thus giving the Latin West, apparently for the first time, a "new logic". The versions of James of Venice served as the vulgate text in the schools until William of Moerbeke revised all the Aristotelian works in the thirteenth century. The versions by James of Venice were obviously closer to the original Greek thought than those versions from the

Arabic, but for this reason Aristotle was less intelligible to the Latins.

Although the basic works of Aristotle appeared in Latin from both Greek and Arabic about the middle of the twelfth century, almost a full century was required before the Latins understood their contents. Until the *Posterior Analytics* was understood, no systematic science of nature, metaphysics or theology could be constructed, for contained therein were the principles of scientific method. Until the *Physics* was understood in its entirety, there could be no clearly defined physical theory, for that was the precise aim of the eight books.

Aristotle's *Posterior Analytics* was widely known in Antiquity and it was generally thought to be his most original contribution to logic. But few ever doubted that it is a very difficult work to understand, even in Greek. By 1159 the *Posterior Analytics* was known to the masters at Paris, but according to John of Salisbury (*d.* 1182) there was scarcely a master willing to expound it because of its extreme subtlety and obscurity: "It contains almost as many stumbling-blocks as it does chapters."[8] John, however, blames this on the bungling mistakes of the scribes and translators as well as on the unfamiliarity of contemporaries with the demonstrative art, so common in geometry and astronomy. The version from the Greek was so terse as to be almost unintelligible; the version from the Arabic was so elaborate that it was vague. John of Salisbury, however, was the first Latin to attempt a brief explanation of its contents

Aristotle's purpose in writing the *Posterior Analytics* had been to solve Plato's dilemma concerning the possibility of acquiring new scientific knowledge. In the *Meno* (80 D-86 D) Socrates proposed to discuss the nature of virtue, a subject about which he admittedly did not have full knowledge. Meno intervenes and objects that all inquiry is impossible, for "a man cannot

[8] John of Salisbury, *Metalogicon*, iv, cap. 6, ed. Webb, p. 171.

inquire either about that which he knows, or about that which he does not know; for if he knows, he has no need to inquire; and if not, he cannot, for he does not know the very subject about which he is to inquire." Plato, through the voice of Socrates, solves the dilemma by means of his doctrine of reminiscence: "All inquiry and all learning is but recollection." Socratic inquiry consists in ordering the questions in such a way as to clear the obstacles to recalling the answer already possessed from a previous existence. This fact offered Plato inescapable proof of the soul's immortality. Thus for Plato the acquisition of new knowledge is really impossible, for it was always actually known; present imprisonment in the body merely makes recollection difficult without a good teacher. The Sophists, on the other hand, proposed a nominalist solution to the dilemma, saying that all learning is simply a collection of individual observations.[9] For them a proof, or "demonstration", is not an intellectual process of perceiving new relationships, but the acquisition of a totally new fact which sheds no light upon other facts. Thus for Plato the answer to a scientific question is no new discovery, for it was always possessed; for the Sophists the answer is a completely new discovery, but it has nothing to do with the question posed.

In answer to the dilemma Aristotle rejected both the Platonic and the nominalist views, retaining the element of truth contained in both. With the nominalists he agreed that the answer to a scientific problem is truly the discovery of something new; but with Plato he agreed that the answer is somehow already known— known *potentially* in the question or problem raised. That is to say, the answer is actually unknown and it must be discovered by scientific investigation, but it is potentially, or virtually contained within the confines of the scientific question. Thus an answer to a scientific

[9] Aristotle, *Post. Anal.*, i, c. 1, 71a 34-b 3.

question is not found in spite of the question, but precisely in terms of the question. The method of scientific investigation, according to Aristotle, consists in the experimental and rational process of beginning with a real problem, and proceeding from the known facts to the solution. This process of analysis may require considerable observation and experimentation, but the solution, to be of scientific value, must be found within the context of the problem and not introduced extraneously. An answer to a mathematical problem, for example, must be found in mathematics, and not in ethics; the solution to a physical problem must be found in physics, not in theology or in metaphysics.

Clearly, the line of demarcation between the proper context of a problem and all that is extraneous depends upon the precise distinction of autonomous sciences. An autonomous science is one which has its own field and principles of investigation, independent of other fields and principles. For Aristotle and many medieval schoolmen there were only four autonomous speculative sciences in the natural order: geometry, arithmetic, natural science and metaphysics. But the number of sciences called "subalternate" were numerous. Astronomy and optics, for example, were said to be subalternate to geometry; music, statics and certain parts of mechanics subalternate to arithmetic; medicine subalternate to natural science. A "subalternate" science is one which has a special field of investigation, but needs the data of a higher science to solve its basic problems, as astronomy needs mathematics. Since the time of Aristotle and the Middle Ages many such sciences have been developed, notably classical and contemporary physics. Of this more will be said later.

The first full-length explanation of the *Posterior Analytics* was written by the renowned Oxford master, Robert Grosseteste, some time between 1200 and 1209. It became the "classical work" on the subject of scientific methodology. The translation of Averroes' great commentary on Aristotle's text, some time around 1220, greatly facilitated the assimilation of this difficult work, and the widely-read commentaries of Albertus Magnus and St. Thomas Aquinas clarified many obscure points. It can be said that the *Posterior Analytics* was more or less understood by everyone from the middle of the thirteenth century until the sixteenth century, and that it was the logical basis for the general physical theory of nature in the Middle Ages. Galileo himself was fully acquainted with the contents of the *Posterior Analytics*, for it was the logical basis of his own theory of nature.

Aristotle's *libri naturales,* or books of natural science, however, did not fare so well in the Latin West. During the first decade of the thirteenth century certain masters at Paris gave lectures based on Aristotle's *Physics*. Having no orthodox guide in the interpretation of these terse writings, Amaury of Bène and his disciple, David of Dinant, utilized John Scotus Erigena's work *On the Division of Nature,* which was a pantheistic version of all the mystic writings of Pseudo-Dionysius. Amaury of Bène and David of Dinant went so far as to identify God with Aristotle's pure potentiality, called first matter. If this were what Aristotle taught, his works on natural science could hardly be sanctioned as textbooks in the schools of Christendom. Consequently, in 1210 the ecclesiastical authorities of Paris condemned Aristotle's books "concerning natural philosophy", and forbade their being taught privately or publicly in the schools under pain of excommunication. This prohibition was repeated by the Papal Legate, Robert Curson, in the Parisian statutes of 1215, the earliest statutes known. The "new logic", however, was required for the mastership in arts. During the twenty years which followed, preachers, theologians and ecclesiastical authorities continued to denounce Aristotle and the natural sciences. Although the Aristotelian books of natural science could not be taught in the schools, individual masters could and did study the forbidden books in private, as their writings testify. In April of 1231 Pope Gregory IX established a commission to examine the Aristotelian *libri naturales* forbidden by the provincial council of Paris twenty-one years earlier. By 1234, it would seem, all the works of Aristotle were permitted in the arts faculty at Paris.[10] About the same time the whole Aristotelian corpus was being taught at Oxford and Toulouse. But it was not until 1255 that the *Physics* and other works of natural science were required for the degree of Master in Arts at Paris.

Masters trained to think within a framework of Christian Platonism could not easily perceive the fundamental ideas of Aristotelian physical theory. The difficult works of Aristotle were often studied with the aid of Avicenna's paraphrase, Ibn Gabirol's questionable *Fount of Life,* and a few minor Greek commentaries translated from the Arabic in the previous century. But the neo-Platonism of Arabian and Jewish philosophers could scarcely help in understanding an entirely different approach to the physical world and the nature of man, an approach which was thoroughly empirical, systematic and naturalistic. The first great scholar to assimilate the Aristotelian science and to advance it by his own research was, without doubt, Albertus Magnus, "a man so superior in every science, that he can fittingly be called the wonder and the miracle of our time."[11] Even in his own lifetime Albert was accepted in the schools as an "authority" on an equal footing with Aristotle. While he was still living he was designated "the Great", and for centuries after his death his name was associated with the wonders of natural

[10] M. Grabmann, *I Divieti Ecclesiastici di Aristotele sotto Innocenzo III e Gregorio IX*, Rome, 1941, pp. 109-10.
[11] Ulrich of Strasbourg, *Summa de Bono,* iv. tr. 3, c. 9.

science, and even with the most improbable scientific
and pseudo-scientific treatises. There were, indeed,
teachers of the Aristotelian books before St. Albert
began publishing his works, teachers such as Robert
Grosseteste, Robert Kilwardby, Adam of Buckfield and
Roger Bacon. But Albert was the first to make the
Aristotelian approach to the physical world his own
and to defend its autonomy against the Platonic
mathematical and metaphysical view formulated by
his contemporaries.

Albertus Magnus of Lauingen was middle-aged when
he first became acquainted with the scientific writings
of Aristotle. He was about forty years old when he
came to "the city of philosophers" around 1240 to
prepare for his mastership in theology. The intellectual
atmosphere of Paris was vastly different from that of
his native Germany, for here he found the wisdom of
the Greeks and Arabs pouring into the intellectual
centre of Christendom. The recent translations of
Michael Scot, particularly of the works of Averroes,
were just reaching the city. Averroes was the most
Aristotelian of the Arabian philosophers, and in his
numerous commentaries (large, medium and small) he
explained the thought of Aristotle word by word. St.
Albert was a prodigious reader with an encyclopaedic
memory. But more than that, he was an acute observer
of natural phenomena with an analytic mind.

Albert was asked by some of his Dominican brethren
to make the *Physics* of Aristotle intelligible to them in
simple language, even though he was a Master in
Theology busily occupied at the University with lectures,
sermons and disputations. Albert acceded to their
wishes, but his plan was far more ambitious than his
brethren could have imagined. Not only did he explain
the general theory of physical science in great detail.
but he intended to explain systematically the whole of
human learning, embracing all the branches of natural

science, logic, mathematics, astronomy, ethics, politics and metaphysics. "Our intention", he said, "is to make all the aforesaid parts [of knowledge] intelligible to the Latins."[12] Strictly speaking, his expositions of Aristotelian thought are neither commentaries nor paraphrases; they are really original works in which the "true view of Peripatetic philosophers" is re-written, erroneous views refuted, new solutions proposed and personal observations incorporated. This, at least, was the opinion of Roger Bacon's contemporaries at Paris, who thought that "now a complete philosophy has been given to the Latins, and composed in the Latin tongue."[13] For this reason, as Bacon himself tells us, Albert's views had as much authority in the schools as those of Aristotle, Avicenna or Averroes. Albert's scientific writings were produced over a period of fifteen to twenty years, during which time he led a very active life. But wherever he travelled, he continued his scholarly and scientific research. "The fruits of his research he passed on to posterity in copious writings, composed with the utmost care, in which he undertook to expound in all its branches nearly every natural science which was known in his time."[14] It came as no surprise to historians of science, when on December 16, 1941, Pope Pius XII declared and constituted St. Albert the Great, Bishop, Confessor and Doctor of the Church, forever the Patron before God of those who cultivate the natural sciences.

The new Aristotelianism of Albertus Magnus was readily accepted by his most gifted disciple, St. Thomas Aquinas, who made it an integral part of his entire thought. Although Aquinas did not search the fields and rivers for new scientific details, he was able to

---

[12] St. Albert, *Lib. I Phys.*, tr.i, cap. 1, ed. Borgnet, iii, 1b-2a.
[13] Roger Bacon, *Opus Tertium*, cap. 9, ed. Brewer, London 1859, p. 30.
[14] Apostolic Letter of 16 Dec. 1941. *Acta Apostolicae Sedis*, xxxiv (1942), 89-90.

formulate the principles and philosophy of Aristotelian science more clearly than Albert. Around 1265 certain masters in the faculty of arts at Paris became obsessed with an "integral Aristotelianism" revealed through the writings of Averroes. Men such as Siger of Brabant and Boethius of Dacia were no fools, but they could not accept the rewritten Aristotelian view of Albert, particularly as it did not agree with Aristotle to the letter. These exaggerated Aristotelians, or "Averroists" as they came to be called, insisted on teaching the eternity of the world, a single intellect for all men and various other points of dubious character, as part of the "pure" doctrine of Aristotle, and therefore as irrefutably true as far as human reason is concerned. This uncritical veneration of "the Philosopher" and adherance to the interpretation of Averroes, even when their views are contrary to Christian faith, aroused the indignation of older theologians who had no sympathy with the "novelties" of Albertus Magnus, Thomas Aquinas or Siger of Brabant. As tension mounted Albert and Thomas penned sharply written treatises against the views of Siger and his followers in an effort to show that neither Siger nor Averroes understood the statements of Aristotle correctly; Averroes, Aquinas claimed, "was not so much a Peripatetic as a perverter of peripatetic philosophy."[15]

It was at this point that Aquinas, at the age of forty-three, felt called upon to write a literal commentary on all the major works of the Stagirite. He asked his friend and confrère, William of Moerbeke, to provide him with a completely reliable, literal translation from the Greek of all the writings of Aristotle and of certain important commentators not yet available in Latin. St. Thomas's purpose in commenting on Aristotle's works of natural science, moral philosophy and metaphysics was to give his contemporaries a scholarly explanation which would

[15] St. Thomas, *De Unitate Intellectus Contra Averroistas*, cap. 2.

be completely faithful to the best translation available and at the same time compatible with the Christian faith as he had expressed it in his theological writings. Wherever Aquinas found a true contradiction or a patent error, he was not slow in finding fault with Aristotle's reasoning or source of information. For this reason the mature commentaries of St. Thomas are extremely valuable for the modern reader; his commentary on the *Physics*, for example, is particularly illuminating as an expression of his complete physical theory of nature.

The Aristotelian Physical Theory of Nature

For Albertus Magnus and Aquinas the eight books of Aristotle's *Physics* were more than just the first of many writings on natural science. The eight books expressed a general, unified theory of all the natural sciences. Aristotle discussed a great diversity of problems in the *Physics*, such as matter and form (bk. i), the concept of nature and causality (bk. ii), motion (bk. iii), place and time (bk. iv), the multiplicity of motions (bks. v-vi) and the unmoved mover (bks. vii-viii). But as an expression of the scientific method expressed in the *Posterior Analytics*, Albertus Magnus and Aquinas saw in the *Physics* a systematic analysis of the most common and fundamental phenomenon observed in nature: physical bodies move. By motion was understood natural, physical change of every kind: coming-to-be, passing-away, increase and decrease, alteration and movement in place. There could hardly be a statement of fact more fundamental than this to express the basic phenomenon common to everything studied in the various natural sciences. Whether the subject matter of study be animate or inanimate, terrestrial or celestial, elementary bodies or highly complex compounds, they all have this in common, that they are physical bodies which in some fashion move. Following the method outlined in the *Posterior Analytics*, Albert and Thomas conceived the basic question of all natural science to be, why do physical bodies move? The common generic subject or "subject-genus"[16] of this science was designated by the general term "physical body capable of movement", or *ens mobile*; its coextensive attribute was expressed by the verb "moves", or simply "motion". Thus Albert and Thomas interpreted the first two books of the *Physics* as defining the common subject matter of all the natural sciences, and the last six books as examining everything necessary to understand the common phenomenon of movement observed in nature.

In this view there is no dichotomy between the study of animate and inanimate bodies, no dichotomy between the study of man and the rest of nature. The activity characteristic of a human being may be vastly different from, let us say, the falling of a stone—that remains to be determined; the undeniable fact for Albert and Aquinas was that animate and inanimate natures are physical bodies manifesting some recognizable kind of temporal change, conditioned by environment and natural dependencies. Recognition of the ultimate unity of all the natural sciences does not imply a denial of the radical difference between individuals, species or genera observed in nature. Albert and Aquinas perceived a profound difference between human nature and the simple elements, but at the same time they acknowledged human nature as part of the totality of physical nature. Ultimately the natural sciences are about the totality of physical nature.

Scientific knowledge, as Aristotle had analyzed it, is not a mere collection of observed facts or true statements. It is the intellectual grasp of the *reason for* the observed fact or true statement. If a statement is not

16 Aristotle, *Post. Anal.*, i, c. 28, 87a 37-38 and i, c. 7, 75a 39-b2.

evidently true in itself, then a man must have some objective reason for holding to it; otherwise the statement is guesswork or prejudice. Observation of facts is absolutely essential to scientific knowledge, but, Aristotle maintained, until a man is struck with puzzlement or wonder about what he sees, he cannot begin to look for rhyme or reason. "We consider ourselves to possess true scientific knowledge of a thing, as opposed to having an opinion, when we think that we know the cause on which the fact depends, directly as the cause of the fact, and that the fact could not be other than it is."[17] Thus, for a general physical theory of nature, it is not sufficient to observe the variety of things in the universe which naturally manifest some kind of movement. The question is, Why does anything move at all? How is it that a stone falls, an acorn grows, ice melts, a bird flies and a man dies? The Aristotelian theory of physical science attempts to answer the basic problem underlying all the branches of natural science: Why does any body move, and move spontaneously in the way we see it to move?

It is interesting to note that in our day many scientists eventually raise problems such as those discussed by Aristotle in his *Physics,* problems concerning change, time, space, continuity, chance and causality. Aristotle preferred to discuss a general theory, involving all such problems, at the very beginning of scientific studies, for three reasons. First, beginners more easily grasp the general theory than the enormous number of details brought to light in a specialized field of study. Second, repetition can be avoided by discussing problems common to many particular branches, if those problems are discussed in general at the beginning. Third, the solution to fundamental problems of physical theory will inevitably influence the solution of special problems which arise in particular branches of investigation.

[17] Aristotle, *Post. Anal.,* i, c. 2, 71b 9-12.

Hence Aristotle's *Physics* was considered the beginners' course in natural science.

For Albertus Magnus and Aquinas, the books of the *Physics* elaborated only the general science of "nature". Other works of Aristotle treated the general structure of the various branches. For example, *De Caelo et Mundo* (*On the Heavens and the Earth*) was considered to be a discussion of bodies which characteristically move by locomotion; this would include the rectilinear motion of elementary bodies and the curvilinear motion of planets, the moon and stars. *De Generatione et Corruptione* (*On Generation and Corruption*) was recognized as a general consideration of chemical alterations, such as are found in the transmutation of elements and the composition of compounds; the *Meteora* (*On Meteors*) was considered a more detailed analysis of such problems. *De Anima* (*On the Soul*) was understood as a preliminary discussion of animate nature; this discussion was continued in the specialized treatises known as *Parva Naturalia* and in the more extensive biological investigation reported in twenty-one books bearing the general title, *On Animals*. Albertus Magnus himself recognized the incompleteness of the Aristotelian corpus.[18] He frequently added entirely new fields of investigation in his rewriting of Aristotle, and sometimes he opened whole new sciences, as in his original research on ore deposits (*De Mineralibus*). But even after all these works of natural science have been studied with the aid of personal observation in the order indicated, St. Thomas recognized that natural science is not complete until the investigator has studied all the species of nature down to the rarest.

Just as in natural things nothing is perfect while it is in potency, but absolutely perfect only when it is in ultimate act—and when it is midway between potency and act, it is somewhat perfect, but not

[18] St. Albert, *Lib. I Phys.*, tr. i, c. 1, ed. Borgnet, iii, 2a.

altogether so—thus it is with science. A science which regards things only in general is not science complete in its ultimate act, but midway between pure potency and ultimate act . . . Hence, it is evident that science, to be complete, must not be content with general knowledge, but must proceed to a knowledge of the species.[19]

In the schools of the Middle Ages the important works were considered to be the *Physics* and *De Anima*. Hence innumerable commentaries on these particular works were composed by medieval masters; they were generally the outcome of classroom lectures delivered to teen-agers. Rarely did the lecturer have the time or inclination to delve into the work *On Animals* or to search the fields, woods and lakes in order to complete his knowledge of natural science. But as Aristotle's *Physics* was almost universally a set book for the degree of Master of Arts, every graduate of a university or *studium artium* could be expected to know something of the general theory of nature. This, however, does not mean that every graduate understood it correctly or completely. Aristotle's *Physics* is a difficult work to comprehend. Albertus Magnus and Thomas Aquinas were both in their forties when they wrote an explanation of the Aristotelian view. Needless to say, their expositions show not only a penetration of the logical coherence of the text, but also mature reflection on the scientific problems involved. Only the briefest summary of their interpretation of Aristotle's physical theory can be presented here.

In the eyes of Albertus Magnus and Aquinas, Aristotle determined the "principles of the subject" of physical theory in the first book of the *Physics,* while in the second he discussed the principles of physical theory itself. The central problem of the first book was to explain the possibility of change, particularly the

19 St. Thomas, *In I Meteor.,* lect. 1, n. 1. Cf. St. Albert, *De Animalibus,* xx, prol., ed. Colon., xii, pp. 1-2.

possibility of fundamental or "substantial" change. The fact of change is abundantly and sometimes painfully evident in human experience: changes of season, birth and death, the escape of all nature into the past. But Parmenides had posed a dilemma which had cast its shadow over the whole development of Greek thought: whence comes the newness of the "changed" body? It cannot have come from nothing, for "nothing" cannot yield anything at all; it cannot have come from the "old" body, for then it would have been there and not there at the same time. Therefore Parmenides was led to maintain that all change is really impossible. The pre-Socratic monists, believing that the universe evolved from one of the elements, were likewise led to deny the reality of change in nature: there is only the appearance of change when the one universal reality assumes different shapes. Plato himself was unable to explain how earthly shadows change while the ideal exemplars remain motionless. The atomists, Democritus and Leucippus, alone were praised by Aristotle for their serious effort to account for the changes evident in experience. But in the last analysis the atomists, too, fail to reconcile theory and experience. According to the atomist theory natural things, such as iron ore, a man or an elephant, are only apparent units; what is called their "change" is nothing more than a local reorganization of ultimate particles, which remain forever immutable, since mutation of ultimate particles, whatever they might be called, cannot be explained. Aristotle's main objection to the atomic theory was that it contradicted human experience. Aristotle, the naturalist, could not favour a theory which turned minerals, plants, animals and men into chimerical unities, one only apparently different from the other. Aware of the dilemma of Parmenides and the efforts of his predecessors, Aristotle claimed to have discovered the only solution which can account for the possibility of intrinsic change of physical substances, whether those substances be men, monkeys or atoms. Without this solution, intrinsic change of physical units is inexplicable.

For Albertus Magnus and Aquinas, Aristotle's originality in physical theory consisted in his discovery of pure potentiality as a reality: a reality which is not pure nothingness and yet not the actual reality we see things to be, a reality which is a pure, passive capacity for change in all bodies. Medieval thinkers before Albert were, indeed, familiar with "first matter", but conceived it as the imprint of imperfection and the finiteness of all creatures, as the chaos from which all creatures are made, and as the innate tendency in creatures to slip back into absolute nothing were it not for the hand of God. This is the neo-Platonic, mystical view widely accepted from St. Augustine's day to our own; but it was not the Aristotelian view of Albertus Magnus and Thomas Aquinas. Medieval thinkers were also familiar with the actuality called "forms", but conceived it as an image of a certain divine perfection added to matter, a luminous entity or a multitude of entities constituting the essential nature of creatures. This view, however, owes more to Ibn Gabirol than to Augustine or Plato; in any case, it was not the Aristotelian view of Albert and Aquinas.

In the Aristotelian view "matter" and "form" are not two things, but two principles of a single, individual thing. One of these principles, namely matter, is the capacity of an individual thing to be what it is, and at the same time to become something else. The other principle is the immediate actuality, or realization of that capacity at any given time, an actuality which makes the individual to exist as a recognizable type of thing. When change takes place, it is not matter which becomes form (as it is sometimes stated in histories of science), nor does one form become a different form.

It is simply the individual *thing* which becomes a
different thing, as hydrogen and oxygen become water.
In Aristotle's view one thing could not *become* anything
else unless there were in that body the ability or
capacity to be something else. It is this capacity which
he called potentiality, or first matter.

We must be careful to understand Aristotle's
explanation correctly, otherwise there is no explanation
at all and we are back with the original difficulty
indicated by Parmenides. Albertus Magnus and
Aquinas never conceived "matter" and "form" as two
separate entities, but solely as *principles* of the existing
thing. The concept of a "principle" in the Greek sense
of the term is extremely difficult to grasp. It is not an
absolute term, but a relative one, so that it cannot in
any way be understood except in terms of what proceeds
from it. The human imagination, however, tends to
spatialize a "principle" of a thing into a thing itself, and
to conceive it as some impish entity inside bodies. This
misconception is not infrequently found in medieval
writers, especially of the fourteenth and fifteenth
centuries. But if matter and form were distinct things,
we should have to ask, "Can they change intrinsically?"
If they cannot, they are no different from the atoms of
Democritus. If they do, from whence came the new
being, from nothing or from the antecedent? How
would this be possible? Avicenna's solution was to
postulate a separate intelligence called "the giver of
forms", which, like Plato's projector, casts shadows
and takes them back. St. Albert insisted that unless
"form" be derived from the potentiality of matter in
the sense that it is simply some actualization induced
by an agent, then all physical change is illusory.

In the eyes of Albert and Aquinas, Aristotle's most
lasting contribution to scientific knowledge was his
distinction between potency and act. By it Aristotle was
able to answer many of the unsolved problems of his

predecessors.[20] By it he was able to explain the possibility and nature of all movement, fundamental changes as well as accidental, natural activities as well as artificial and unexpected ones. The first book of the *Physics* did nothing more than establish an hypothesis which could explain the observed fact of change; it rested with the particular natural sciences to see the verification of this hypothesis.

St. Thomas saw in the second book of the *Physics* an exposition of the basic "principles of natural science", or physical theory. The natural sciences are not concerned with every type of change indiscriminately. They are not directly concerned, for example, with the building of ships or the production of beer; these changes are induced by human artisans, and the proper technique belongs to the art concerned. But all artisans must work with natural materials of some sort and accept the conditions of nature; these materials and conditions are of direct concern to the natural scientist. Likewise, chance occurrences and the abnormal can be understood only in terms of what is regular and normal. Therefore, the domain of the natural sciences, for Aristotle, embraced everything that is produced by nature and functions in accord with its innate capacities for better or for worse. In Greek thought the term "nature" was technically defined as a principle, or source, directly responsible for movement toward and possession of a commensurate state of being.[21] Aristotle, however, showed in the first book of the *Physics* that there are two vastly different types of principle which can be responsible for movement, namely "form", the active source of characteristic activities, and "matter", the passive capacity for accepting change. Thus the goal

[20] St. Albert, *Lib. I Phys.*, tr. iii, cap. 15; St. Thomas, *In I Phys.*, lect. 9, n. 3.
[21] Aristotle, *Phys.*, ii, c. 1, 192b 21-3; see J. A. Weisheipl, *Nature and Gravitation*, River Forest, 1955, pp. 1-32.

of the natural sciences is to understand the source of characteristic activities and the native passivities of all natural things. This source, or nature, cannot, of course, be understood except through detailed observation, which alone renders the idea of "source" meaningful.

The natural scientist wants a detailed understanding of physical reality as it actually exists. For Aristotle this meant understanding physical bodies actually endowed with sensible characteristics such as colour, texture, shape, weight, odour and sound; it meant understanding a particular species animate or inanimate, in all its physical conditions, dependencies and contingencies. Every physical being, for example, depends upon some agency to bring it into existence, as water needs heat to become vapour and an infant needs parents to conceive it. Further, every species in nature tends toward a determined state of completion and maturity; this determined state is technically called the final cause of its movement and development. Therefore, the natural scientist must examine the efficient, final, material and formal causes responsible for natural phenomena and species before he can claim to have scientific knowledge of nature.

Aristotle pointed out that the naturalist's approach to nature is vastly different from the mathematician's approach to the same nature. The mathematician, as will be explained more fully in the following section, must leave out of consideration whatever is not amenable to quantification, measurement and mathematical operations. Obviously, matter and form, nature as a source of motion, efficient and final causality, are not amenable to measurement; hence these principles have no meaning to a mathematician approaching the physical world. Even motion, time and the so-called secondary sense qualities (colour, heat, sound, taste and odour) have no meaning to the mathematician approaching nature, except as measured quantities; but these idealizations are not the motion, time or sense qualities of human experience. The peculiarity of mathematical abstraction of every kind, as we shall see, is that it must consider quantities and measures as though they had an ideal existence distinct from tangible matter and actual movement.[22] The characteristic of natural science, on the other hand, is that it accepts all the data of human experience and tries to interpret the full data according to the given natural principles of activity (form) and passivity (matter), efficient causality and purpose.

Since "nature" is defined as an active or passive source of motion, no physical theory can be complete unless it determines the full meaning of motion, or change.[23] Albertus Magnus and Aquinas considered the remaining books of the *Physics* as a comprehensive analysis of motion, i.e., change, the commensurate attribute[24] of nature as a principle. The unique reality of motion in itself, as Aristotle recognized, is difficult to comprehend, because it is at one and the same time an actuality and a capacity for more of the same. The actuality which characterizes motion is radically different from every stable act which might be called "form"; and the potentiality proper to motion is vastly different from the permanent capacity called "matter". The fleeting reality of motion exists only when and where the body is moving; it is not at all something stretched out between two points as though it were a line. Motion is peculiar in that it is a process which simultaneously has accomplished something and still needs to go on. Aristotle's view was succinctly expressed in the scholastic axiom: Everything which is being moved, will be moved. As soon as the need to go on

[22] Aristotle, *Phys.*, ii, c. 2, 193b 31-35.
[23] Aristotle, *Phys.*, iii, c. 1, 200b 12-15.
[24] St. Thomas, *In III Phys.*, lect. 1, n. 1; see Aristotle, *Post. Anal.*, i, c. 4, 73b 27-74a 3.

ceases by nature or through violence, motion ceases. For this reason Aristotle defined the process of moving as "the activity of a body in potency, precisely as it is a capacity for more of the same act" [25] He knew of no other way to define motion, without involving the thing to be defined in the definition.

Aristotle's definition, it is true, implies the notion of infinity, but St. Thomas thought this was as it should be. In Greek thought "infinity" signified that which is necessarily without a final determination in the sense that its potentialities for further determination are never exhausted; there always remains something more to be determined.[26] Infinity, therefore, had meaning only in the context of potentiality, a potentiality which always has room for more act, like a line which always remains divisible. (For this reason some of the Greek Fathers of the Church explicitly denied infinity as an attribute of God.) In Greek usage, an actual infinite, or an infinite with final determination in act, would have been a contradiction in terms. With regard to the definition of motion, St. Thomas admitted that the phrase "in potency precisely as it is a capacity for more of the same act" does express infinite potentiality; this is essential to the nature of motion. In itself motion is an infinite expectancy of further motion; conversely, when there is no expectancy, there is no longer motion. The termination of motion comes not from motion itself, but from the nature of natural bodies, which always act for an end and which require movement as a means of attaining that end—for example, a heavy body falling freely to its natural place.

In the fourth book of the *Physics* Aristotle discussed place and time, because all motion necessarily exists in place and time. Unlike the modern idea of undifferentiated space, place was conceived as the environment

[25] Aristotle, *Phys.*, iii, 201a 10-11.
[26] Aristotle, *Phys.*, iii, c. 6, 207a 1-2.

into which and out of which physical bodies move. One type of motion is defined entirely in terms of place; this is called locomotion. Other motions, such as generation, corruption, alteration and growth, not only occur at a given place, but are very much affected by the environment. In Aristotle's view, place is both qualitative and quantitative; as an environment it is qualitative, but as absolutely or relatively stationary with respect to determined points it is quantitative and similar to space.

Physical motions also occur in time. Some motions, like locomotion, alteration and growth, require time and are measured by the length of time required. Fundamental changes, like coming-into-being and passing-away, take place in time — they obviously do not take place in eternity. However, these changes cannot be measured by time in the sense that coming-into-being would require a certain length of time, since a new thing either has or has not come into being. Therefore all fundamental changes are said to be instantaneous.

The reality of time (not to be confused with that of some measure of time) is extremely difficult to grasp, as anyone who considers the matter will discover. Time, like motion, is a reality in flux. Yesterday does not exist and tomorrow never comes; all that exists is the now, which never stays. Yet, no one would say that the indivisible "now" is what men call time, for time is universally considered a certain interval between two points of time, as for example, one second, an hour, a day, a year. Aristotle considered past events to be past for everything in the universe, and the future to be future for everyone, even for a man sitting on the outermost sphere. This means that past, present and future are the universal measure for all events, and the measure of events taking place right now. Aristotle, therefore, defined time as the universal measure of all motion, or events precisely as those events have a "before" and "after", a past and future. But if past and future do not exist, how can there be measurement of motion unless the mind is actually thinking of past and present? Aristotle himself raised the question, "Supposing there were no mind existing, would there still be time?"[27] Aristotle's reply is admittedly obscure, but he seemed to suggest that the measuring is actually in the mind, and if there were no mind, only measurable motion would exist. This at least was the interpretation of Alexander of Aphrodisias, Galen, Themistius, Simplicius and Averroes. Albertus Magnus, however, dismissed the obscure passage and, after a lengthy discussion, rejected outright the universally accepted view of his predecessors. For Albert, time is as much a physical reality as motion, independent of any mind measuring. The past is past even if no one thinks about it. Just as motion exists in a body only when and where it is moving over a distance, so time exists in the universal motion of the whole. Just as the reality of motion is not solidified between two points, so the reality of time cannot be solidified between two points. If a whole day existed simultaneously, it would no longer be a day. It belongs to the reality of time to be flowing and passing; but in so passing it divides past and future. The young St. Thomas accepted the traditional view expressed by Averroes in this matter, but in his maturity he followed the lead of Albertus Magnus.

In the fifth and sixth books of the *Physics* Aristotle discussed the qualitative and quantitative diversity of motions. Qualitatively, fundamental changes are radically different from secondary (accidental) changes of place, size and sense of characteristics. Secondary changes, which Aristotle called "motions" in the strict sense, all presuppose a fully constituted natural subject acting and reacting in accord with its nature. Locomotion, alteration, growth and decrease are recognizably distinct movements of physical bodies. But in funda-

[27] Aristotle, *Phys.*, iv, c. 14, 223a 21-2.

mental changes there is no true subject at all, only the purely passive principle accepting the activity of efficient agencies. Having nothing, not even its own existence, "first matter" cannot be called a true subject of change. Despite the difference between these types of true motion, they can all be defined as some kind of "activity of a body in potency, precisely as it is a capacity for more of the same activity". Clearly, such a process could not be confused with any or all of the final states acquired through motion, for every final state, whether it be place, quality or quantity, is necessarily static.

Motions can also be analysed on the basis of the quantities involved, namely the moving body, the distance traversed and the length of time required. That a motion be numerically one, a single body must traverse one distance once. But a plurality of any one factor makes the motion numerically different. Thus a single body may move the same distance at different times, or it may pause at intervals during one length of time; or many bodies may cover the same distance at the same time or in different times. For Aristotle motion is not a quantity; it is an activity indirectly quantified by reason of the extended body which traverses a magnitude in a given time. Motion, however, is essentially a continuous process. But if motion were not in an extended body traversing a distance in time, motion could not in any way be continuous. Therefore the continuous character of dimensional quantity is responsible for the continuity essential to motion, but it is not identical with it. Motion, for Aristotle, always remains a process, an activity of a body in unrequited capacity for more of the same. Such a process is by no means self-explanatory or self-sufficient. In a physical theory of nature this process needs to be explained in terms of the causes responsible for it.

In the last two books of the *Physics* Aristotle raised the final question necessary for a complete physical theory of nature. This question concerned the efficient causes of motion, ultimately the efficient cause of all observed motions. Summarizing the thought of Aristotle, St. Thomas noted that change or motion for its own sake is an impossibility, because motion is essentially a means of acquiring some end.[28] The entire *raison d'être* of motion is derived from the purpose of the motion, and this implies an agency capable of producing the motion and attaining the end. The agency responsible for producing many motions observed in experience is immediately recognized; for example, the boy throwing the stone, the archer shooting the arrow, hydraulic pressure conveying water through pipes or lifting weights. In all such cases of mechanical, or "unnatural", motion, we can easily recognize the cause which satisfactorily explains the particular motion. Other types of motion, such as the animate movement of animals, can be traced to a true power within the moving body itself. By definition a living thing is a physical organism capable of moving itself. Thus a worm is the cause of its own wiggling, and the canary of its own flying; we have no need to look beyond the animal to explain vital activity. But for inanimate natural motions the cause of motion is not so easily recognized. "It is in these cases that difficulty is experienced in deciding whence the motion is derived, e.g., in the case of light and heavy bodies. . . . It is impossible to say that their motion is derived from themselves, for this is a characteristic of life and peculiar to living things."[29]

All natural motions, it is true, somehow originate from the natural body itself, for "nature" was defined earlier in the *Physics* as the source directly responsible for movement. But this "nature" in the case of inani-

[28] St. Thomas, *QQ. Disp. De Pot.*, q. 5, a. 5; *Contra Gent.*, iii, c. 23.
[29] Aristotle, *Phys.*, viii, c. 4, 255a 1-6.

mate bodies cannot be called a true *cause* of movement, for the inanimate body obviously does not move itself. Therefore these automatic motions originating from a nature already constituted have their cause in something antecedent to the nature, the cause which originally produced the nature. Thus the constituted nature acts now as the instrument of the agency originally responsible for that nature. Simply expressed, this means that every nature must be constituted in existence by the actualization of pure potentiality, first matter. Once a particular nature has been brought into existence, the body moves spontaneously in accord with that nature. In inanimate bodies such a nature can never be a self-moving cause, for then they would hardly be called inanimate. Even living things must be generated; once generated, a living thing is a true efficient cause of certain movements, of others it is merely an instrument acting in accord with an implanted nature.

From this consideration St. Thomas concluded that every natural body is generated and hence "moved" in the sense that every individual is brought into existence. Although Aristotle imagined the celestial bodies and matter to be eternal, he argued that the eternal movement of those celestial bodies could not have been produced by the physical bodies themselves. Therefore everything in nature *is moved*. But, clearly, nothing which is moved can move itself into act, for then it would have to be in act before being in act. An infinite regression in agencies or the aggregate of natural agencies does not explain the totality of physical bodies moved. To say that there is no cause would be to deny the reality of existing motions in the universe. Therefore there must exist some cause of motion and the coming-into-being of all nature. This mover, if it is to explain the evidence of human experience, must be totally different from the world of bodies and motion. That is to say, the cause of all physical bodies and their evident movement must itself be unchangeable, immovable and totally separated from matter.

Only one point need be added to this summary of Aristotle's *Physics*. Albertus Magnus and St. Thomas never confused Aristotelian physical theory with metaphysics. The unexpected discovery, that there must exist a being separated from matter in order to explain observed motion and change, presented them with an entirely new field of study, a study which Aristotle called "first philosophy", or metaphysics. Thus there would be no metaphysics, if there were no beings entirely immaterial and immovable — or if we did not know about them. That there are such beings (at least one) cannot be known except through a physical theory which attempts to explain the motion and changes actually existing in nature. Aristotle expressed his own view concerning the relation of physics to metaphysics when he said,

> If there were no substance other than those which are formed by nature, the science of nature would be the first science; but if there is an immovable substance, the science of this must take priority, and for this reason be called "first" philosophy, and hence universal, because it is first.[30]

This view was strongly defended by Albertus Magnus and Aquinas against those who would consider physical theory and natural science a branch of metaphysics. The autonomy of physical theory and the natural sciences rested on the validity of natural principles in explaining natural phenomena.

PHYSICAL THEORY AND MATHEMATICS

A more subtle problem developed, however, concerning the mathematical sciences and a purely physical theory of nature. Plato, as we have already seen, not

[30] Aristotle, *Metaphysics*, vi, c. 1, 1026a 28-31.

only considered mathematics the highest knowledge, but also the key to understanding all reality. For this reason he urged Eudoxus to discover the geometrical form which could explain all the known movements of the heavens. For Plato and his disciples, the physical structure of bodies and motion was derived from the geometrical structure of the eternal idea, and this in turn was derived from its arithmetical structure. This desire to see in mathematics the ultimate explanation of physical phenomena had a long history and may be termed the mathematical approach to nature.

Aristotle never denied the validity of pure mathematics, nor the utility of mathematics for the study of nature. He himself was abreast of the mathematics of his day. But he showed no little impatience with his contemporaries, Platonists and Pythagoreans, for whom philosophy had become identical with mathematics, and who expended their energies searching for the number out of which the world is composed.[31] Aristotle claimed that they had given up the search for the cause of perceptible things and "say nothing of the cause from which change takes its start". For him, to fail to explain movement through all the causes given in experience is to annihilate the whole study of nature. Apparently, Aristotle considered the use of mathematics in astronomy, optics, harmonics and mechanics, a profitable first step in understanding natural phenomena. But there remained the more important task of examining the physical natures themselves and of discovering all the causes involved in natural phenomena. It was this task which Aristotle felt was being neglected by his contemporaries.

In the thirteenth century Albertus Magnus showed the same impatience toward his contemporaries who denied the autonomy of the natural sciences, and who reduced natural science to mathematics, and mathematics to

[31] Aristotle, *Metaphysics,* i, c. 8-9.

metaphysics.[32] Albertus Magnus fully understood the "error of Plato" as explained by Aristotle in various parts of the *Physics* and *Metaphysics,* and he neglected no opportunity to reject it forcefully. Not only did Albert deny that universal natures are subsistent and separated from tangible matter—for this was commonly rejected in the thirteenth century—but he called "absurd and entirely false" the claim that the principles of natural phenomena are mathematical. He denied that natural phenomena could be explained entirely in terms of mathematical structure, and that mathematical principles could be derived from metaphysical unity and plurality. Without identifying them, Albert noted that among Latin authors there were certain "friends of Plato" who subordinated natural science to mathematics, thereby denying its rightful autonomy in the solution of natural problems; but they also subordinated mathematics to metaphysics, thereby denying the autonomy of mathematics as well.

Among Albert's unnamed contemporaries who could be called "friends of Plato", one might single out Grosseteste, pseudo-Grosseteste, Kilwardby and Roger Bacon. These men, of course, never thought of themselves as Platonists. In fact, they were among the earliest masters who lectured on the Aristotelian books in the schools. However, they promulgated the very view Albertus Magnus attacked throughout his writings.

Robert Grosseteste, later Bishop of Lincoln and first Chancellor of the University of Oxford, had considerable influence on the development of English thought in the thirteenth and fourteenth centuries. As the first instructor of the Oxford Franciscans, he was held in high esteem by the Order for many generations. Although he never became a Franciscan, his own approach to Sacred Scripture, theology, metaphysics and the natural sciences

---

[32] J. A. Weisheipl, "Albertus Magnus and the Oxford Platonists," *Proc. Am. Cath. Phil. Assoc.,* vol. 32, 1958, pp. 124-39.

fitted well the spirit of Greyfriars. In the history of science Grosseteste is best known today for his "metaphysics of light", a view which was largely inspired by Sacred Scripture, Pseudo-Dionysius, St. Augustine and Plato.[33] In this view God is the primordial light which shone in the darkness of first matter, giving it the first form of substantial being. The union of this luminous form and chaotic matter constitutes a created substance, which, like an indivisible point, cannot be seen with human eyes, because it is inextended in space. Such is the created substance of angels and the human soul. But the luminous form of corporeal creatures radiates by self-multiplication in three dimensions, generating the form of corporeity. Corporeal creatures are, therefore, constituted of chaotic matter, substantial form (light) and the form of corporeity, which gives mass and extension to a physical body. Further self-multiplication of the luminous rays begets subsequent forms, both essential and accidental. The rays of all light, according to Grosseteste, follow determined laws of mathematical proportionality. The mathematical laws governing self-multiplication of visible light rays, their reflection, refraction and the like, also regulate the constitution of bodies and the activities of spiritual light. Hence for Grosseteste the key to understanding all reality lies in geometrical optics.

In his commentary on the *Posterior Analytics*, which was written before 1210, Grosseteste gave much consideration to the application of mathematics to visible, physical phenomena, particularly the application of mathematics to visible light rays. He recognized, as did Albert after him, that optical phenomena depend upon two factors: the mathematical law and the nature of light. But for Grosseteste the mathematical law alone could give a true understanding, a proper explanation of

---

[33] A. C. Crombie, *Robert Grosseteste and the Origins of Experimental Science,* Oxford, 1953, pp. 91-134.

the visible phenomena; not even his "metaphysical theory of light" provided him with a proper explanation in optics. In the technical terminology of Grosseteste, a demonstration involving the nature of light as the middle term, could convince one of the mere fact (*quia*) of the phenomenon, but not a proper (*propter quid*) explanation for it; only a middle term which is mathematical could provide such an explanation for the visible phenomenon.[34] These technical expressions had been explained by Aristotle in the *Posterior Analytics* and were understood by all scholars from the thirteenth to the sixteenth century; the crux of the controversy concerning mathematics and physical theory lay, as we shall see, in the kind of demonstration, or knowledge, mathematics provides in the realm of nature.

Some years later Grosseteste's "metaphysics of light" was developed in great detail by an anonymous author of a *Summary of Philosophy*, written around 1263. For many centuries this vastly erudite work was attributed to Grosseteste himself, because of the obvious heritage. The procession of translucent light from its primordial source through the hierarchy of forms generated by self-multiplication down to the specific composition of natures was presented with elaborate care. Problems of astronomy, psychology, biology and alchemy were discussed with the aid of elementary geometry and some principles of optics. But the approach to all these problems was patently neo-Platonic, pseudo-mathematical and at times mystical. The author is clearly an Englishman, who has no sympathy for the novelties then being introduced by Albert and Aquinas.

Of far greater importance is the view of Robert Kilwardby, who as a secular master in arts lectured for many years at the University of Paris, probably from 1237-45. His monumental treatise, *De Ortu Scientiarum*

---

[34] Grosseteste, *Comm. in lib. Post.*, i, c. 8, Venice edition of 1552, fol. 9rb-va.

(*On the Origin of Sciences*), seems to have been composed at Oxford some time around 1250, after he had entered the Dominican Order. In sixty-seven chapters Kilwardby explained the nature, origin and division of all the departments of human knowledge from metaphysics to the seven mechanical arts. While discussing the relation of mathematics to the natural sciences (chap. 25), he presented an elaborate theory of intellectual abstraction, which, while it has some foundation in Aristotle, is in fact neo-Platonic and Arabic in origin. Briefly, Kilwardby conceived the three main branches of speculative knowledge (natural science, mathematics and metaphysics) as rooted in three grades or forms found in all physical reality: perceptible qualities, dimensive quantity, metaphysical substance. The domain of natural science extends to all perceptible, extended substances, but it "abstracts" from the individual characteristics. Mathematics further "abstracts" from perceptible qualities and studies only extended substances. Finally metaphysics "abstracts" from extension and contemplates the naked substance in all its purity. The term "abstraction" signifies the mental process by which certain aspects of reality are left out of consideration, while that which is abstracted is retained for consideration. In the "three degrees of abstraction", mentioned by Kilwardby, the process is one of sloughing off outer shells, one at a time, preserving the inner shells and fruit. The natural scientist peels off the shell of individuality in his consideration of physical phenomena; the mathematician peels off the shell of sense qualities; the metaphysician peels off the shells of dimensionality and number, discovering the fruit of pure being.

The interesting point in Kilwardby's theory is that all perceptible qualities and observable phenomena are rooted in the dimensionality of bodies. Therefore the proper explation of physical phenomena is to be found in mathematics; the science of nature is technically "subalternated" to geometry. But dimensions, for Kilwardby as for Plato, are derived from a determined number, just as a triangle is derived from the number "three". Therefore the principles and theorems of geometry are proved in algebra; the science of geometry is "subalternated" to the science of numbers. But the root of number is unity, the unity of intelligible substance. Therefore the laws governing numerical plurality, proportionality and infinity are explained by metaphysics, the philosophy of pure substance and of the First Cause of all plurality. In this view metaphysics alone is an independent and autonomous science, because it alone studies the highest cause of all reality. The natural sciences are insufficient to disentangle the complex problems which confront the thoughtful observer of nature; mathematics and finally metaphysics are required.

Kilwardby's brilliant treatise, which is still unpublished, undoubtedly contains the most thorough and vivid account of the mathematical theory of nature produced in the thirteenth century. An introduction to all the human arts and sciences, it brings divergent sources together in a synthesis which is oblivious of different "schools" of thought. Kilwardby, guided by his Arabic sources, could see no real divergence between the teaching of Aristotle and that of Plato. The historian of medieval thought, however, can immediately recognize the unconscious Platonic and neo-Platonic structure of Kilwardby's theory of physical science. Although Kilwardby penned the most lucid theoretical justification of the mathematical approach to nature, he wrote nothing concerning mathematics, optics or astronomy. As far as one can judge from his writings, he was not particularly interested in mathematics or in problems of natural science.

Kilwardby's most illustrious disciple, Roger Bacon, took up the banner of mathematics against what he called the contemptible ignorance of his contemporaries, although he himself made no contribution to this science.[35] Later centuries, relying on legend and imagination, fashioned for themselves an idol, a prophet of science and a champion of free thought; and it is difficult to separate this idol from the thirteenth-century philosopher, mathematician, linguist and scientist who became a Franciscan friar late in life. Even one who reads his writings finds difficulty in distinguishing an outrageous boast from historical fact. Unfortunately many important writings of Bacon, dating from his professorship at Paris as a secular master, have not yet been discovered, writings to which he referred in later years. The three famous *Opera,* which are invariably cited in any discussion of Friar Bacon, were written late in life under difficult circumstances. The bitter invectives, irresponsible charges and vociferous contempt which veil a secretiveness born of many disappointments are perhaps not a safe index to his true scientific knowledge. But the historian must form a tentative judgement on the basis of available evidence.

The works of Roger Bacon abound with praise and appreciation for mathematics, languages and experimentation, just as they abound with derision and contempt for his contemporaries and predecessors who claimed to know these subjects, but were in fact ignorant of them. He dismissed all existing translations of Aristotle as completely unreliable, and he charged William of Moerbeke, Michael Scot and Gerard of Cremona with not understanding "either the sciences or the tongues". He frequently singled out ignorance of mathematics as the cause for the universal decline of learning in the thirteenth century. Writing to Clement IV between 1266-8, Friar Bacon declared:

The neglect of [mathematics] for thirty or forty

---

[35] D. E. Smith, "Place of Roger Bacon in the History of Mathematics," *Roger Bacon (Commemorative Essays),* Oxford, 1914, pp. 176-7, 182-3.

years has nearly destroyed the entire learning of Latin Christendom. For he who does not know mathematics cannot know any of the other sciences; what is more, he cannot discover his own ignorance or find its proper remedies. So it is that the knowledge of this science prepares the mind and elevates it to a well-authenticated knowledge of all things. For without mathematics neither what is antecedent nor consequent to it can be known; they perfect and regulate the former, and dispose and prepare the way for that which succeeds.[36]

Among the few who passed Bacon's quill unscathed was his hero and model, Robert Grosseteste, whom Bacon most probably never met in this life.

The influence of Grosseteste on Bacon's philosophy of science is unmistakable. Bacon conceived the three speculative sciences (natural science, mathematics and metaphysics) as corresponding to a real hierarchy of forms in nature. Unlike Kilwardby, he did not emphasize the supreme role of metaphysics in human knowledge. Rather, he followed Grosseteste in according mathematics the magical power of opening all doors. This view is clearly expressed in the above quotation, where Bacon states, "Without mathematics neither what is antecedent nor consequent to it can be known". The natural sciences are antecedent to mathematics in the hierarchy of knowledge, and metaphysics is consequent. Clarifying the function of the mathematical sciences, Bacon explains, "They perfect and regulate the former, and dispose and prepare the way for that which succeeds". Mathematics perfects natural sciences by giving an ultimate explanation of natural phenomena, and regulates them by judging the utility and validity of experimentation in each branch of natural science. Because mathematics alone could perfect and regulate the natural sciences, Roger Bacon was convinced that

[36] Bacon, *Opus Maius*, pt. iv, dist. 1, c. 1, ed. Bridges, Oxford 1897, i, pp. 97-8.

"it is impossible to know the things of this world [*huius mundi*], unless one knows mathematics."[37] In confirmation of this view, Bacon, like his predecessors, could point to the role of mathematics in astronomy and optics.

Lest one should gather the impression from what has been said, that Grosseteste, Pseudo-Grosseteste, Kilwardby and Bacon were voices crying in the wilderness, we must hasten to add that theirs was the common voice among men who thought at all of a philosophy of the physical sciences. The hands were indeed those of Aristotle, but the voice was unconsciously that of Christian Platonism, because such was the living tradition of Western thought, confirmed by the twelfth-century translations from the Arabic. What is more, it was a voice which did not disturb the meditation of theologians. Undoubtedly there were not many writers in the mid-thirteenth century who insisted on the importance of mathematics, as did Grosseteste and Bacon. But such an insistence could hardly disturb the theologians of the day. Theologians could appreciate the "three degrees of abstraction" ascending to God through the hierarchy of forms in nature; they could even nod amiably at the thought of God, the author of mathematics. Roger Bacon might be exceptionally Aristotelian in his desire for experimentation, but theologians could continue to meditate serenely upon the writings of St. Augustine.

The real innovation in Western thought came from the pen of Albertus Magnus, who, although he was himself a theologian, introduced certain philosophical views which appeared strange to his contemporaries. The theological implications of Albert's philosophy became disturbingly clear in the writings of his disciple, St. Thomas Aquinas. During his own lifetime, Aquinas' views were not infrequently associated with certain heretical doctrines. However, the evident sanctity and

[37] Bacon, *Opus Maius*, pt. iv, dist. 2, c. 1, *ed. cit.*, i, p. 109.

intellectual stature of these two men staved off personal condemnation for views repeatedly shown to be orthodox. But in one of the unfortunate blunders of history, some of those views were implicated in the episcopal condemnations of Averroism. On the third anniversary of Aquinas' death, 1277, Tempier, Bishop of Paris, condemned a list of two hundred and nineteen propositions; eleven days later Kilwardby, Archbishop of Canterbury, condemned a shorter list at Oxford. The propositions which affected the doctrine of St. Thomas were officially revoked on February 14, 1325, two years after his canonization.

In the new Aristotelianism of Albert and Aquinas what place did mathematics occupy in the halls of thought? What kind of assistance could mathematics give to the solution of physical problems? What kind of scientific "explanation" did they think mathematics could offer for natural phenomena? The view of Albertus Magnus and Aquinas was essentially that of Aristotle, but greatly refined and more precise. Scientific knowledge had developed considerably since the days of Aristotle: Euclid had systematized geometry, Archimedes discovered important laws in hydrostatics, Ptolemy devised a highly satisfactory system of the universe and contributed greatly to the development of optics. These developments encouraged and apparently confirmed the Platonic theoreticians of knowledge in the thirteenth century. The new Aristotelianism of Albert and Aquinas took issue with the Platonic view of mathematics and the use of mathematics.

The Platonists considered mathematical structure as an objective form antecedent to physical characteristics and phenomena. But for Albert the "structure" which mathematics studies is a mental abstraction, an idealized conception formed by leaving out of consideration all the complications of matter, and the actual flow of movement and time. Hence, for Albert, mathematical structures are not antecedent, but consequent upon the physical constitution of nature. In other words, the "nature" of any existing thing is determined to inform a more or less determined quantity of matter, so that an elephant is not as small as a mouse; it is determined to express itself in a determined figure, so that the figure of a woman is not mistaken for that of a pumpkin. Albertus Magnus considered these dimensions and figures to be the direct concern of the natural scientist; they do not antecede any physical nature, but are the quantitative characteristics expressive of a type. Physically existing dimensions and figures are not "mathematical" quantities, but the original source of these idealized versions of certain elementary dimensions and figures. In this idealization the mathematician eliminates from consideration not only irregularities found in nature, but the very nature begetting physical quantity. Consequently, the mathematician has eliminated from consideration the fundamental problem of how one type of body changes into another, the difficult problems of purpose, natural agency, activity, movement and temporality. In other words, the astonishingly real characteristics of nature perceived in human experience elude static idealization. Even when the mathematician applies his principles and ingenuity to movement and time, as in astronomy, physical movement and time have already flown from the linear co-ordinates, and all that is left is an idealized distance, static and serene.

The Platonists saw in mathematics a deeper explanation of physical phenomena, an insight more divine than that afforded the natural scientist who examines qualitative and contingent characteristics. For Albertus Magnus, however, the insight afforded by mathematics is not deeper, but more superficial; the "explanation" proffered by a mathematical approach to nature is not an *explanation,* but a suitable *description* of observed natural phenomena, a description, it is true, which may help in discovering an explanation. Mathematical abstraction, for Albert, necessarily eliminated from consideration the four types of natural causation (efficient, final, formal and material); what it retains is a shadow reflecting something of the "formal" cause. The shadow, or quantitative image, such as figure, measure, number and velocity, which is utilized in a mathematical approach is, therefore, not an "explanation" of why events take place, but measured data which can be accounted for in terms of geometrical figures and determined proportions.

In confirmation of their view Albertus Magnus and Thomas Aquinas frequently alluded to the example of Ptolemaic astronomy, which was accepted by everyone in the Middle Ages as a practical and theoretically satisfactory account of all known movements in the heavens. Ptolemy's complicated system of eccentrics and epicycles, representing the finest in Greek mathematical theory, was able to account for all the known facts; it was able to "save all the appearances" (*salvare apparentias*), but it did not attempt to explain why heavenly bodies move at all. As Pierre Duhem, the noted French historian of science, has shown, the attempt "to save the appearances" has played a dominant role in the history of astronomy and physical theory.[38] The phrase, however, is somewhat misleading in English; it means accounting for all the known facts. The medieval schoolmen understood perfectly well that Ptolemy's theory was a geometrical construction devised to fit the observed facts, and did not imply a belief in phsyically existing spheres, eccentrics and epicycles. Hence the schoolmen found no embarrassment in teaching the doctrines of Aristotle and Ptolemy side by side. St. Thomas, discussing the difference between the Aristo-

[38] P. Duhem, " $E\dot{\omega}\zeta\epsilon\iota\nu$ $\tau\dot{\alpha}$ $\Phi\alpha\iota\nu\dot{\omega}\mu\epsilon\nu\alpha$. Essai sur la notion de théorie physique de Platon à Galilée", *Annales de phil. chrétienne*, vol. vi, 1908.

telian and Ptolemaic systems, noted that after the time of Aristotle, "Hipparchus and Ptolemy devised [*adinvenerunt*] the motions of eccentrics and epicycles in order to account for those facts of celestial motion which were evident to the senses; hence this is not a proven theory, but a kind of supposition."[39] Following Simplicius, Aquinas noted that the visible motions of the celestial bodies "are produced either by the motion of the object seen or by the motion of the observer . . . it makes no difference which is moving"[40]; but if both the heavens and the earth are considered at rest, there can be no accounting for the apparent motions. Hence, the Ptolemaic theory was considered a satisfactory mathematical description, but not a physical, causal explanation of celestial motions.

Astronomy, as we have seen, was universally considered an intermediate science (*scientia media*) between mathematics and the science of nature; technically, it was said to be "subalternate" to geometry, because it used geometrical principles to account for natural phenomena. We have likewise seen that for Grosseteste and his disciples mathematics could not only describe the appearances, but could give the ultimate reason for those appearances. In other words, Grosseteste claimed that the underlying mathematical structure of natural phenomena was causally responsible for the phenomena, and hence mathematics alone could give an explanation of the visible world. For this reason he insisted that intermediate sciences, such as astronomy and optics, attained the true and proper (*propter quid*) explanation of natural events.

For Albertus Magnus and Aquinas, on the other hand, the natural sciences constitute an autonomous field of investigation, having their own method and principles of research; the principles of natural science are all those

[39] St. Thomas, *In I De Caelo*, lect. 3, n. 7.
[40] St. Thomas, *In II De Caelo*, lect. 11, n. 2 and lect. 12, n. 4.

factors in nature responsible for events as they exist and as they are encountered in experience. The mathematician's approach to these events is the approach of an outsider: he is equipped to give a considered opinion, but the decision rests with the household. For Aquinas, the contribution of mathematics to the solution of a problem in natural science is like the contribution of metaphysics to a legal case, or like the contribution of theology to a discussion of poetry. They can all say something illuminating about the remote factors involved, but they are out of touch with the immediate factors, which must eventually settle the case. For this reason St. Thomas remarked that an explanation of natural events through mathematical principles is equivalent to an explanation through a "remote cause".[41] The quantitative measure of motion, time and bodies necessarily abstracts from the physical event, hence it is "remote"; it describes real quantities, and hence it is a formalized copy, or "cause" of the event. The peculiar character of mathematical abstraction led both Albert and Aquinas to insist on the propaedeutical role of mathematics, including astronomy, optics and mechanics. After these introductory disciplines of the *Quadrivium* have been mastered, one ought to get on with the task of investigating the real problems of nature, beginning with the basic problems touching the whole of physical science and ending with detailed investigation of the individual species of both living and non-living things.

DEVELOPMENTS IN THE LATER MIDDLE AGES

"If we had to assign a date for the birth of modern science," Pierre Duhem once remarked, "we should undoubtedly choose the year 1277, when the Bishop of Paris solemnly proclaimed that there could exist many

[41] St. Thomas, *In I Post. Anal.*, lect. 25, n. 6.

worlds, and that the ensemble of celestial spheres could, without contradiction, be moved in a straight line."[42] Whatever may be the propriety of assigning a date of ecclesiastical significance, it is clear that after the episcopal action of 1277 theologians and philosophers alike felt justified in rejecting indiscriminately the new views of Albertus Magnus and Thomas Aquinas, who were unjustly implicated in the episcopal condemnation. However, rather than being a victory for any "party", or even a defeat for Aristotle, Albert or Aquinas, it merely encouraged individualism and universal ecclecticism. Constantine Michalski has somewhat more disparagingly described the aftermath as a growth of various currents of criticism and scepticism.[43] But "criticism and scepticism" is perhaps too strong and misleading a description of a highly complex period. Aristotle and all of Greek and Arabic science had come to stay, but the interpretation of ancient thought varied considerably among the ever increasing number of masters and universities. The rapid growth of provincial universities, bulging libraries, new towns and trade routes made the acquisition of profound learning increasingly more difficult. Late medieval thought became increasingly more concerned with linguistic and logical analysis of problems and statements; indeed, not infrequently the original problem got lost in the subtlety of analysis.

One significant innovation of the early fourteenth century was brought about by the English Franciscan, William of Ockham (c. 1284-1349), who is currently one of the contenders for the international title, "Father of Modern Science".[44] Before his untimely departure from

[42] P. Duhem, *Études sur Léonard de Vinci*, Paris, 1906-13, vol. ii, p. 412.

[43] C. Michalski, *Bulletin de l'Académie polonaise des sciences et des lettres (classe d'hist. et de phil.)*, Cracow, 1924-7.

[44] E. Whittaker, *Space and Spirit*, Edinburgh, 1946, pp. 43-53, 139-43; *From Euclid to Eddington*, Cambridge, 1949, pp. 65-7; *History of the Theories of Aether and Electricity*, Edinburgh, 1951, vol. i, p. 3.

the University of Oxford, the "Venerable Inceptor", as he came to be called by his disciples, lectured on the *Sentences* of Peter Lombard some time around 1318. This commentary is without doubt the most important of Ockham's writings, not only because it was the basis for the formal charge of heresy made against him by the Chancellor of the University, but also because it contained all the essential philosophical and scientific ideas defended by the "founder of nominalism". Later works, such as the numerous treatises on logic, the two commentaries on the *Physics* and minor theological works, added nothing significant to the principles expounded by Ockham while he was bachelor of theology, not yet thirty-five years of age. After his excommunication in 1328 until his unreconciled death twenty-one years later, Ockham's time was chiefly taken up by political controversies and what were claimed to be the heresies of Pope John XXII.

Ockham's nominalism, or terminism (as his disciples preferred to call the new system), was intended to embrace the whole of human knowledge, but it particularly affected logic, natural science and theology. However, not everyone who accepted Ockham's views in logic felt inclined to go all the way with him in applying nominalist principles to the natural sciences. Hence it is advisable to distinguish Ockham's logic from his theory of physical science, although Ockham frequently insisted that his view alone was the true teaching of Aristotle in both disciplines.

The expressions "nominalism" and "terminism" have their origin in Ockham's teaching concerning logical supposition, that is, the reality "supposed", or represented by the use of nouns (*nomina*) or noun terms, derived from verbs, adverbs, conjunctions, prepositions and syncategorematic terms.[45] Earlier schoolmen had

---

[45] Cf. C. Michalski, "Le Criticisme et le scepticisme dans la philosophie du xive siècle", Cracow (reprint 1926), pp. 78-80;

taught that only terms used as the subject of a proposition could properly have supposition; and depending upon the context, the subject term could refer either to a material expression, as a written word, or to a particular individual, or to an objective common nature shared by many individuals. For Ockham every part of speech could be a noun, and hence every term in a sentence had some supposition. But Ockham denied the reality of an objective "common nature"; for him that which is called a "common nature" is nothing more than a mental fiction (*intentio animae*).[46] The "fiction of abstract nouns" derived from other parts of speech was considered by Ockham to be the cause of many difficulties and errors, since beginners commonly imagine that distinct nouns correspond to distinct realities. These abstract nouns are used in speech for the sake of brevity, and other parts of speech are employed for the sake of "elegance in discourse". That which is objectively real is not the mental fiction, but the individual things capable of independent existence; these independent individuals Ockham called "absolute things" (*res absolutae*). Hence, although supposition was extended to every term in a sentence, the reality described by those terms was restricted to individual absolute things. The crux of Ockham's natural philosophy, however, lay in what he admitted as an "absolute thing".

The basic principles of Ockham's theory of physical science is that only *res absolutae* are existing realities; everything else is a name, a term describing its condition briefly or elegantly. But Ockham admitted only two kinds of *res absolutae*: substances and qualities. "Besides absolute things, namely substances and qualities, nothing is imaginable, either actually nor potentially."[47] Only

---

P. Vignaux, "Nominalisme", in *Dict. de théol. cath.,* xi, cols. 733-42.

[46] Ockham, *Sent.,* i, dist. ii, qq. 4-8.

[47] Ockham, *Summa Logicae,* pt. i, cap. 49.

individual sense qualities (such as colour, heat, shape and heaviness), individual matter and individual forms are objectively real, for they are "absolute things" each capable of independent existence. All other terms, such as quantity, motion, time, place, velocity and causality, refer to nothing real over and above the individual substance with its qualities.

The word "quantity", for example, was considered a mere abstract noun by Ockham because that which men call "quantity" is not an absolute thing. It cannot exist by itself; moreover, it can "increase and decrease" and even disappear altogether by the absolute power of God without affecting the individual substance or its qualities. This is proved by the phenomenon of condensation and the mystery of the Eucharist. In condensation and rarefaction — a favourite problem for Ockham — the volume of space occupied by a body changes without affecting the quantity of matter, substantial form or any of the absolute qualities. By his absolute power God could even condense the whole universe so as to eliminate all extension, thus making all the distinct parts of the universe to exist in one point of space.[48] In fact, God does just this, according to Ockham, in the Holy Eucharist, where the distinct parts of Christ's body and blood (i.e., matter, form and qualities) are completely preserved without any extension in space. Hence quantity is nothing real over and above substance and qualities, which naturally ought to occupy space, but need not: in a condensed body they occupy little space, and in the Eucharist they occupy no space. Such theological reasoning led Ockham to conclude that when we speak of the "quantity" of matter in the sense of its mass or size, we resort to the expedient of using the simple term "quantity" or its equivalent to discuss individual matter, form or qualities, which ought to occupy space but need not, and when they do occupy space,

[48] Ockham, *Sent.*, iv, q. 4 H and q. 6 J.

the amount could be otherwise. Thus for Ockham quantity is not a physical reality, but an abstract noun employed for the sake of brevity and without any determinable significance. For this reason Ockham reduced all problems of physical quantity to problems of grammar and logic.

More significant, and more frequently discussed, is Ockham's view concerning motion. In the natural sciences many terms are used to designate a body not at rest, terms such as movement, motion, change, growth, condensation, heating, freezing and many others derived from verbs. But all such terms are fictitious abstract nouns designating no reality other than matter, form or sense qualities. Nothing besides absolute things, namely substances and qualities, can exist; but what is termed "movement" is clearly not a substance or a quality, because it cannot exist independently of other things. Moreover, if it were an "absolute thing", God could conserve motion without ever giving it a termination; but clearly this is self-contradictory. Therefore "motion" and similar tems are mental fictions designating nothing but the individual matter, form or quality. For Ockham the concept of motion is constructed of two elements, one positive, the other negative. The positive element is the individual body or quality; this was intended by Aristotle when he defined motion as "an actuality of a body". The negative element is the negation of rest; Aristotle expressed this when he added to the definition "in potency". The noun "local motion", therefore, briefly expresses the complex phrase, "a body which was in one place, and later will be in another place, in such a way that at no time does it rest in any place".[49] Clearly, the negative element of this definition cannot exist outside the mind; only the individual body, which is identical with its termination, or rest, can conceivably be given real existence by God.

[49] Ockham, *Sent.*, ii, q. 9, and *Tract. de Successivis*, tr. i, c. 2.

Hence the term "motion" designates no reality other than an individual body negatively described as not being at rest. From this Ockham concluded that the search for a "cause of motion" is entirely superfluous, since motion is nothing over and above the individual physical body.

Some authors have seen in Ockham's view of motion a foreshadowing of the modern theory of relativity.[50] The physicist, however, who considers relativity to be objective might feel self-conscious in the company of nominalists. In any case the purpose of Ockham's so-called relativism was to deny that motion, velocity, distance, time, position and the like are in any way determinable realities over and above the individual matter, form or quality. In the Eucharist the body and blood of Christ are present without motion, time or confinement in place; therefore none of these can be realities over and above the *res absolutae*. The young Franciscan departed from his predecessors for whom motion was a process, and he took a path far different from that of his contemporary Mertonians for whom motion was a measurable quality (see below, p. 76). His path was the "modern way" of grammar and logic. But there can be no doubt that his chosen torch was the light of Christian faith; in accord with the Franciscan spirit Ockham wielded even grammar and logic in the service of theology. Like the modern logical positivist, Ockham had at least the merit of emphasizing the importance of verbal analysis. But fortunately this was not universally considered the best way of solving problems of physical science.

The views of the Venerable Inceptor spread quickly among logicians and theologians both in England and on the Continent. But French and Italian masters in

[50] E. J. Dijksterhuis, *Die Mechanisierung des Weltbildes,* Berlin-Heidelberg, 1956, p. 198; E. Whittaker, *Space and Spirit,* pp. 139-43.

arts were, for the most part, too Aristotelian (some would say, too Averroist) to accept Ockham's nominalism in natural science. Parisian masters, like John Buridan, would not have proposed a new theory of "impetus" to explain projectile motion, had they considered motion a fiction needing no causal explanation. Not infrequently modern historians of scientific thought present Ockham's nominalist theory of motion as a necessary first step in the discovery of impetus, the law of inertia, the various Mertonian laws of motion and the Copernican system.[51] However, Ockham might have considered this a somewhat dubious honour.

When P. Duhem discovered the theory of "impetus" in the writings of John Buridan and his disciples, he felt that at long last he had found "the Parisian precursors of Galileo". Duhem maintained that it was the overthrow of the Aristotelian distinction between natural and compulsory motion by means of the theory of impetus which led directly to the principle of inertia, the corner-stone of modern physics. This thesis, which has found favour among many historians of modern science, may now be considered an unfortunate hypothesis formulated by a great pioneer in the history of medieval science.[52] Buridan's theory of impetus is, in fact, a natural development of Aristotelian physical theory, and it is very different from the principle of inertia proposed in the seventeenth century.

Aristotle had described natural movement as the characteristic activity arising spontaneously from the nature of a physical body, if it is not impeded by circumstances; a stone spontaneously falls to the earth, if

[51] E. A. Moody, "Galileo and Avempace", in *Journal of the History of Ideas,* vol. xii (1951), pp. 397-410; "Ockham and Aegidius of Rome", *Franciscan Studies,* vol. ix (1949), pp. 417-38; E. Whittaker, *loc. cit.*

[52] A. Maier, *Studien zur Naturphilosophie der Spätscholastik,* ii. Rome, 1951, pp. 113-314; J. A. Weisheipl, *Nature and Gravitation,* pp. 33-64.

it is not sustained. Compulsory motion, on the other hand, was described as an unnatural activity foisted upon a natural body, contrary to its innate tendencies: such would be the upward movement of a stone. We have already seen that Aristotle found no difficulty in recognizing the agency responsible for compulsory motion: it is whoever or whatever imposes the force (cf. above pp. 46-8). But Aristotle did experience difficulty in explaining how this unnatural force propelled the projectile after it has left the impelling agency. That the force must remain *foreign* throughout the flight is clear, for otherwise it would become native and natural, thus offering no explanation for the observed difference between natural and compulsory motions. Moreover, experience shows that all foreign forces are gradually diminished and eventually overcome by the innate tendencies. In order to maintain the alien character of such compelling forces, Aristotle considered it necessary to locate them in a medium, such as air or water, which could propel the projectile until the forces diminished in power. Valid objections were raised against this theory by the sixth-century Aristotelian, John Philoponus of Alexandria, but his commentary on the *Physics* was not available to the Latin West.

It is possible, as Michalski and Maier have noted, that the Franciscan contemporary of Ockham, Francis de Marchia, was the first Latin author to raise serious objections to Aristotle's explanation and to propose an alien source of motion within the projectile. While lecturing on the *Sentences* at Paris (1319-20), the Franciscan bachelor of theology discussed the instrumental causality of the sacraments, explaining how the sacraments could contain the grace they confer. One minor objection to such a notion of instrumental causality was theoretically raised, namely the case of projectiles which contain no internal instrumental power. Instead of answering the objection in the manner of his predecessors, Marchia took occasion to discuss Aristotle's explanation of projectile motion. He insisted that no way can be found to explain such motion, except by positing a "power impressed on the body by the mover", a power called *impetus*. Like the force given to a hammer in use, the impressed power is alien, because it does not belong radically to the instrument. After a relatively short time, the impressed power is dissipated and the body is free from constraint. Marchia also explained how the celestial spheres could revolve by themselves "for a time" because of an impressed force. Although Marchia's digression added nothing to the accepted view of sacramental causality, it was discussed briefly by his disciples and successors in the Franciscan chair of theology at Paris.

Some years later the problem was discussed more thoroughly and resolved more satisfactorily by John Buridan, who lectured on the Aristotelian books of natural science approximately from 1325 until 1358. We do not know when Buridan first discussed the problem of impetus, nor do we know whether he conceived the theory independently of Marchia, whom he could not have heard in the schools. In any case, this "celebrated philosopher" from Béthune frequently discussed the problem of projectiles, rejected Aristotle's view and clearly conceived "impetus" as a qualitative power given to the projectile by the original mover. For Buridan impetus was still an alien, accidental, unnatural power of movement; hence, the Aristotelian distinction between natural and violent motions was preserved. But more satisfactorily than Marchia, Buridan maintained that impetus is overcome only by contrary natural forces of the body which from the very beginning of motion tend to destroy the unnatural situation. Buridan suggested that the eternal motion of the spheres might be explained by a single impetus, since the celestial bodies have no innate contrary motion, according to factual experience and the teaching of Aristotle. He also attempted to give a physical explanation of natural acceleration on the basis of increments of impetus produced by the natural gravity of heavy bodies. All of Buridan's efforts were directed toward expanding and correcting the Aristotelian theory of physical science. Rather than overthrowing the Aristotelian distinction between natural and compulsory motion by means of the theory of impetus, Buridan proposed the theory in order to preserve it.

Buridan's disciples, Albert of Saxony, Nicholas Oresme and even Marsilius of Inghen, accepted the theory of impetus to explain the continuation of motion in a body separated from the mover, and to preserve the distinction between natural and violent motions in the universe. This became the common "Aristotelian" teaching throughout the fifteenth and sixteenth centuries; some even anachronistically claimed it as "the opinion of St. Thomas". However, it must be noted in passing, that sometimes impetus was referred to as a "mover accompanying the body", or as the "efficient cause of motion". If such expressions are taken literally, they are scarcely compatible with Aristotle's distinction between animate and inanimate nature, described above. On this point Francis de Marchia was more consistent with Aristotelian principles when he explained impetus as an instrument of the agency responsible for the impact.

Before considering the position of Galileo in the development of scientific theory, a very different innovation in fourteenth-century science merits careful attention. This significant innovation was a new mathematical approach to nature, quite independent of the Platonic ventures of the thirteenth century. It was inaugurated by Thomas Bradwardine (c. 1295-1349), who, like Isaac Newton three centuries later, was more devoted to problems of theology than to natural science.

The theological renown of Bradwardine, one-time Fellow of Merton College and later Archbishop of Canterbury, has been immortalized by Chaucer in *The Nun's Priest's Tale*. Bradwardine not only led a full academic life, but he subsequently became prominent in affairs of Church and State.

Considering the age in which he lived, there can be little doubt that Bradwardine was a mathematical genius, who, in the apt words of Anneliese Maier, "would have wanted to write the *Principia mathematica philosophiae naturalis* of his century".[53] Leibniz, the great mathematician and philospher, might have said with more historical accuracy that Thomas Bradwardine (rather than his disciple, Richard Swineshead) was "the first to introduce mathematics into scholastic philosophy".[54] In a burst of enthusiasm which reminds us of Roger Bacon and Galileo, Bradwardine declared, "It is [mathematics] which reveals every genuine truth, for it knows every hidden secret, and bears the key to every subtlety of letters; whoever, then, has the effrontery to study physics while neglecting mathematics, should know from the start that he will never make his entry through the portals of wisdom."[55] Contemporary historians of medieval science refer to Bradwardine with some justification as the "founder of the Merton School" of science, just as P. Duhem referred to Buridan as the founder of a new school of scientific thought.

The importance of Bradwardine in the development of physical theory rests largely on his influential *Treatise on the Proportion of Velocities in Moving Bodies*, composed in 1328, while a student of theology at Oxford. By way of apology Bradwardine declared his purpose in writing the treatise:

Since every successive motion is commensurate to

[53] A. Maier, *Studien*, i, Rome, 1949, p. 86, n. 10.
[54] Letter to Thomas Smith (1696).
[55] Bradwardine, *Tractatus de Continuo*, MS. Erfurt, Amplon. Q. 385, fol. 31v.

another with regard to velocity, natural philosophy, which studies motion, ought not to ignore the proportion of motions and velocities; and since an understanding of this is necessary and extremely difficult, and not discussed fully in any part of philosophy, we have accordingly composed the following work on the proportion of velocities in moving bodies.

The purpose of the treatise was to propose a mathematical law of dynamics universally valid for all changes of velocity in local motion. In the first three chapters the "true" law governing rectilinear motion was presented; in the last chapter the way was prepared for its application to the rotational motion of the spheres.

With regard to motion Aristotle had rightly observed (*Phys.*, vii, 4-5) that nothing can be moved unless the power of the mover is greater than the resistance; the velocity of motion (which is the distance traversed in a given time) depends in some way on the excess of moving power overcoming the resistance. For Aristotle, velocity was considered a property of all movement, but he insisted that not all types of movement are comparable in velocity, because of the equivocal use of the term "distance". Thus Aristotle would not say that burning is more rapid than walking, or that the sun moves more quickly than a boy matures, because the so-called distance covered in local motion, growth and change of quality is not identical. The time intervals can be compared because time is identical for all physical motion, but the "distances" cannot be compared. In Aristotle's view motions of one single kind, such as all rectilinear motions, can be compared in velocity. And since velocity depends upon the excess of the moving power over resistance, one can say that a mover twice as powerful as another will move a body twice as far in the same time (or the same distance in half the time) as the other moving the same body in the same time. It does not follow, however, that a mover one-half the strength of another will move half the distance in the same time or the same distance in twice the time; a mover half the strength of another may be unable to overcome the resistance, and hence not move at all.

Bradwardine found much wanting in this traditional formulation of the law, namely that twice the velocity follows from doubling the moving power or from halving the resistance. It was valid for a particular case, but as a universal law of dynamics it was "insufficient and based on a false assumption": insufficient, because the law holds only when the ratio of moving power to resistance is 2:1; based on a false assumption, because it follows from this that any mover can move anything, while in fact, motion follows only if the power of the mover is greater than the resistance. In other words, the traditional formula was not theoretically valid for all cases, and it did not preclude the possibility of the moving power's becoming less than the resistance. Therefore Bradwardine proposed a new law, maintaining that a double velocity must follow from the entire power-to-resistance-ratio *duplata,* i.e., not multiplied by two, but squared, for twice a *proportio tripla* is one which contains the ratio 3:1 twice. But only the square of the ratio, i.e., only 9:1, contains twice the *proportio tripla,* for the ratio 9:1 is composed of 9/3 . 3/1. Thus only the ratio squared can give twice the velocity. Similarly, triple velocity follows from the ratio *triplata,* i.e., raised to the third power. Conversely, half the velocity would follow from the *medietas* of the ratio, that is, the square root of the ratio; one-third the velocity follows from the cube root, and so forth; thus the mover can never be less than the resistance. Bradwardine's exponential function was theoretically valid for all cases, and it eliminated the possibility of zero velocity. Today this latter situation is taken care of by a logarithmic function.

In order to approach this problem Bradwardine automatically eliminated the concept of motion as a process, an "activity of a body in potency", substituting for it the concept of motion as a velocity, which is a ratio of distance to time. Bradwardine conceived velocity as an inhering, qualitative ratio which could be intensified and remitted like any other quality, but always according to his laws of proportionality. This was more pointedly expressed by an anonymous author, possibly Richard Swineshead, when he defined motion through its four causes:

The material cause of motion is whatever is acquired through motion; the formal cause is a certain transmutation conjoined with time; the efficient cause is a proportion of greater inequality of the moving power over resistance; and the final cause is the goal intended.[56]

Bradwardine's identification of motion and velocity was the first step necessary in the mathematization of motion, since a process, as such, is impervious to mathematical treatment. This conception of motion as a real ratio produced by and proportioned to another real ratio is not only different from that of Aristotle, but it is clearly opposed to that of Ockham.

Bradwardine, however, was not content with a mathematical function for terrestrial, rectilinear velocities; he wished to discover a single kinematics, if not a dynamics, for both terrestrial and celestial motion. The fourth chapter of Bradwardine's treatise was devoted to the special problem of rotational motion in order to establish the commensurability of celestial and terrestrial velocities. Aristotle had maintained that the two velocities could not be compared, any more than a circle and a line could be compared. Bradwardine's task was to discover some common basis whereby the two veloci-

[56] *Tractatus de Motu Locali Difformi,* Cambridge, Gonville & Caius MS. 499/268, fol. 212ra.

ties might be commensurate. In effect he discovered what everyone takes for granted today, namely that a circle can be imagined as stretched out into a straight line; this is what we do when we measure the circumference of a circle. After rejecting three opinions concerning the comparison of rectilinear and rotational velocities, Bradwardine proposed his own solution: since the velocity of any motion consists in a body's traversal of a given fixed space (real or imaginary) in a given time, it seems reasonable to say that "the velocity of a rotating sphere is measured by the velocity of the most swiftly moving point of the sphere". The "fixed space" would be the linear distance the fastest moving point would traverse, if the point were going in a straight line. In the case of the celestial spheres, velocity would be determined by the distance that the fastest moving point, e.g., a planet, would cover in a given time, if it were moving in a straight line instead of in a circle by reason of the sphere. This at least allowed Bradwardine to give some meaning to a comparison of terrestrial rectilinear motion and the rotational movement of the spheres. Bradwardine would indeed have wanted to write the *Principia Mathematica* of his century. But history had to wait for Isaac Newton.

Bradwardine's treatise became a best-seller of the fourteenth century; it was immediately acclaimed throughout Europe and before long it became part of the university course at Oxford, Paris, Vienna, Prague, Padua and Florence. The mathematically satisfactory function discovered by Bradwardine incited younger men to discover some way of reducing all changes to a single law of dynamics. This was of particular concern to the young masters in arts at Oxford, many of whom were Fellows of Merton College in the generation after Bradwardine. Among the most eminent Fellows of Merton who devoted considerable attention to the mathematical calculations were William

Heytesbury, John Dumbleton and Richard Swineshead.

Authors after Bradwardine generally divided discussions of motion into two parts: dynamics, or the relation of mover to resistance, and kinematics, or the relation of distance to time. The Mertonians were especially interested in solving the problems of kinematics. Basically the Mertonian problems were twofold: (i) to find the proper function, or latitude which determines different types of motion; and (ii) to determine the equivalence of accelerated motions in order to render such motions intelligible and mathematically useful. Bradwardine employed a geometrical manner of presentation in all of his major writings, even in theology. His disciples, however, did not emulate the clarity of this Euclidian procedure. Instead, they were fond of the tedious "letter-calculus", the use of letters of the alphabet to represent ideas, which facilitated academic disputations, but which was detested by humanists and almost unintelligible to later readers.

The first of the Mertonian problems concerned the function according to which velocity can be determined, that is, the *distance* or *latitude* determining the velocity of various changes of quality, condensation, rarefaction and local motion. Aristotle had insisted that the velocity of such motions could not be compared because the type of distance in each case is unique. The Mertonians had to determine, first of all, some characteristic factor which can be acquired or lost for any particular quality; temperature, for example, would be such a characteristic for heat. This was called the "latitude" of a quality. Then the latitude had to be conceived as having "degrees" before an increase or decrease of that quality could be compared to local motion. Thus the degrees of latitude acquired in a given time would determine the velocity of that alteration. Theoretically, if one could determine the velocity or acceleration of motion (i.e., "motion according to its effect"), one should be in a position to determine according to Bradwardine's law the proportion of mover to resistance (i.e., "motion according to its cause "). The kinematic problems alone proved too difficult even for the "Calculator", Richard Swineshead, so that the fundamental problems of dynamics remained untouched. Swineshead, however, laid the foundation for the development of infinitesimal calculus; Leibniz, one of the discoverers of calculus, had such respect for Swineshead's *Book of Calculations* that he wished to republish the work.

The second fundamental problem for the Mertonians was to give meaning to the concept of accelerated motion. The meaning of uniform motion is clear enough: it is the covering of equal distance in equal time. In uniform motion the "degree of velocity" remains constant and no velocity or "latitude of motion" is acquired or lost. But in accelerated or decelerated motion "the latitude of motion is increased or decreased", as Dumbleton expressed it. Thus the "quality of motion", or the "degree of velocity" at any given instant of acceleration, cannot be the ratio of the whole distance to the time, but it must be a special velocity at that instant. But what is the meaning of "velocity at that instant" when velocity has always meant using the linear quantities of distance and time? It is one thing to talk about the quantity of the whole motion, that is, the total velocity (*velocitas totalis*) of distance covered uniformly in a given time; but it is a different thing altogether to speak about the quality or intensity of motion at any given instant (*velocitas instantanea*), much less to assign a value to that degree. Bradwardine himself had made the distinction between the "quantity of motion", meaning the total velocity, and the "quality of motion", meaning the velocity at any given instant. But it remained to later Mertonians to determine the meaning and the value of that instantaneous velocity.

As for the meaning of instantaneous velocity, Heytesbury and all those after him had agreed that it is the
total velocity which the body would have if it were to
move uniformly with the constant velocity possessed at
that instant, e.g., at any given moment a body is said to
be travelling so many miles an hour, even though it is
not moving with constant speed. This is really an application of Bradwardine's conclusion in the fourth chapter to the case of accelerated motion at any instant:
instantaneous velocity is equivalent to "the line described
by the fastest moving point". There was no difficulty
about this. But it does not follow from this that the
total velocity of uniformly accelerated motion is equivalent to the instantaneous velocity at the last instant,
i.e., the total distance covered is not equivalent in value
to that covered at the maximum velocity were it to be
constant for an equal length of time; nor is the total
velocity equivalent to just any instantaneous velocity.
The problem was to determine to which instantaneous
velocity of uniformly accelerated motion the total velocity is equivalent. The Mertonians unanimously agreed
that the total velocity of uniformly accelerated motion
must be equivalent to the mid-point, or to the average,
of the accelerated motion from its starting point to the
maximum grade. This "Merton rule of mean speed",
as it has come to be known, was geometrically demonstrated in various ways, but the clearest and best known
in the days of Galileo was devised by Oresme at Paris.

The significance of Bradwardine's physical theory was
that it inaugurated a new mathematical approach to
nature, an approach which gained momentum as it
spread from Oxford to continental universities, to
Galileo, to Newton and the highly complex physics of
our own day. The Mertonian treatises were not commentaries on Aristotle, but highly sophisticated works
dealing with logical analysis and the "calculations of
motion", including local motion, alteration (intension

and remission) and augmentation, which invariably meant condensation and rarefaction. These writings of early fourteenth-century Oxford quickly spread throughout Europe; they were read, copied, explained in the schools and improved in detail. The Mertonian vocabulary was the basis for the seventeenth-century scientific terminology. But Bradwardine's goal, namely to embrace all motions, terrestrial and celestial, and all variations of velocity, in a single mathematical formula, was not even reached by Galileo; it had to wait for Newton to find temporary realization. Bradwardine's conception of motion as a real geometrical structure (velocity) functionally dependent upon another geometrical structure (the ratio of mover to resistance) was not fully understood by his successors, who generally incorporated it into the Aristotelian *Physics,* particularly on the Continent. The clumsy terminology, the awkward letter-calculus and the geometrical demonstrations were not sufficient to distinguish a new mathematical approach to nature from the Arisotelian physical theory. The tension was most keenly felt around the time of Galileo, but the occasion was the suitability of accepting the Copernican theory of planetary motion.

Galileo, it must be remembered, was educated at Pisa, where he was taught the Aristotelian books of natural science, the astronomy of Ptolemy and the now traditional "calculations of motion". But he was deeply impressed by current developments in pure mechanics, particularly by the work of Giovanni Benedetti and Simon Stevin (who actually did drop weights from a height, only to find that all landed simultaneously). In pure mechanics Galileo was able to contribute to the almost universal development. The question of whether the acceleration of freely falling bodies was directly proportional to the distance or to the time, had been debated in the schools since the fourteenth century. With his predecessors Galileo agreed that the speed could not be proportional to the time simply; but he found a satisfactory figure when he multiplied the distance by the time squared. He disregarded questions concerning the nature of "impetus" and the relation of impressed force to the mass of a body; Galileo considered the total quantity produced by mass times velocity as a single force called *impeto*. Thus for Galileo the *impeto*, or *momento*, was not a quality by which compulsory motion takes place, as Buridan had conceived impetus, but the quantity of motion measured by the mass of a body times its velocity (mv). It was commonly held in Galileo's day that no quantity of motion, or momentum, could ever be lost. From this it logically followed that "any velocity once received by a body is perpetually maintained as long as the external causes of acceleration or retardation are removed, a condition which is found only on horizontal planes."[57] Therefore Galileo was not concerned with explaining the existence of motion or its nature, but only with the change or cessation of motion. For him it was not the cause of motion or its continuation which needed to be explained, but change of direction and velocity.

Most of Galileo's energies were devoted to "demonstrating" the truth of the Copernican system against both the defenders of Aristotle and the defenders of Ptolemy. His battle was against two views which had coexisted happily for centuries in every university of Christendom. If Galileo had argued with the Aristotelians on the grounds of natural philosophy, or if he had argued with the defenders of the Ptolemaic system on grounds of pure astronomy, probably there would have been no Galileo legend to haunt the textbooks of modern science. Instead, Galileo proposed a new science which would supplant both natural philosophy and astronomy. For him the truth of this new science

[57] Galileo, *Discourses on the Two New Sciences*, Third Day, prob. ix, prop. 23.

depended upon the irrefutable truth of the Copernican system, which he had "demonstrated" as a fact of nature.

Galileo was fully aware that the principal aim of astronomers in the past was to account for the knowr facts of celestial motions: to save the appearances.[58] But he was not satisfied with this. "Although it satisfies an Astronomer merely Arithmetical", Galileo insisted, "yet it does not afford satisfaction or content to the Astronomer Philosophical."[59] He refused to accept Osiander's statement in the preface to *De Revolutionibus Orbium* that this system was simply a mathematical device which accounted for the known facts better and more simply than did the Ptolemaic theory. Galileo insisted that the Copernican account of celestial spheres carrying planets like our earth around the sun with perfect regularity was a literally true account of nature, and that no other account could be true. He was willing to venture into Biblical exegesis in order to maintain the literal truth of Copernicus' undeniably advantageous theory of planetary motion. Patience and caution were never outstanding traits in Galileo's character; and his unhappy venture into the domain of theology provoked ecclesiastical authorities to take an unfortunate stand in an unsatisfactory case. Galileo had not, in fact, *demonstrated* the truth of the Copernican theory; today no one accepts Galileo's "demonstrations" on behalf of the heliostatic system. The Holy Office, likewise, had insufficient grounds for rejecting the system.

Galileo's originality, however, did not lie in his "proofs" of the Copernican system. Nor did it lie in the discovery of certain laws of mechanics, important as these were. Nor did it lie in an "overthrow of Aristotle" by means of the telescope or falling weights. It

[58] Galileo, *Dialogue on the Great World Systems,* Third Day, ed. G. de Santillana, pp. 349-50.
[59] Ibid., p. 350.

lay in his insistence that the book of nature is written *only* in mathematical language. "This book is written in the mathematical language, and the symbols are triangles, circles, and other geometrical figures, without whose help it is impossible to comprehend a single word of it; without which one wanders in vain through a dark labyrinth."[60] These sentiments were, of course, shared by Johann Kepler, but Galileo was less Platonic in the presentation of his physical theory than his German contemporary.[61] Galileo deliberately set aside any search for the "causes" of gravity and acceleration, insisting that he wished to describe mathematically how bodies do fall, the path described by a projectile, the actual parallelogram of forces and the movement of stars. However, Galileo never considered his mathematical description of nature to be a mere accounting for the known facts, a mere "saving of the appearances"; it was for him the only true *demonstration* of the way nature acts. By "demonstration" Galileo understood a true, necessary and proper (*propter quid*) explanation, as described by Aristotle in the *Posterior Analytics*. In other words, Galileo considered a strictly natural science of nature to be incapable of offering such an explanation for natural events. For him "substantial form", "gravity", "sense qualities", "nature", "final cause" and the like were mere words which explain nothing. For this reason he found fault with certain parts of William Gilbert's otherwise admirable work *On the Magnet*. For this reason he objected to the Aristotelian approach to natural science. Geometrical demonstrations alone could explain the true operations of nature. Whatever could not be caught in mathematical abstraction, such as secondary sense qualities, essences and causes, were either subjective or did not exist for Galileo. For this

[60] Galileo, *Il Saggiatore*, Florence edition of 1842, p. 171.
[61] Cf. E. A. Burtt, *The Metaphysical Foundations of Modern Science*, London, 1932, pp. 52-71.

reason he felt keenly about the absolute truth of the Copernican theory; it was the framework of his new science replacing Aristotelian physical theory and the traditional view of astronomy.

Whether Galileo's strict mathematical realism would be amenable to the complexities of modern quantum and relativity physics might be questioned. Isaac Newton's view of a "mathematical way" which allows room for another kind of investigation into the causes of gravitation, was more consistent with the traditional views of human knowledge. Newton was convinced that nature acted as though bodies attract each other in direct proportion to the mass and inversely proportionate to the distance; but he knew that mathematics was incompetent to determine the cause of gravitation. Newton's universal mathematical mechanics for terrestrial and celestial motions reached the goal seen by Thomas Bradwardine, and at the same time it preserved the traditional structure of a true "middle science" (*scientia media*) between pure mathematics and natural science. Rather than eliminate a strictly natural science of the universe, as Galileo had done, Newton's mathematical principles of natural philosophy allowed freedom for an autonomous natural philosophy seeking to solve its own problems with its own natural principles.

CONCLUSION

We have traced in miniature certain developments in physical theory from the early age of the Christian Fathers through the Latin Middle Ages to the age of Galileo and modern classical mechanics. We have seen that the Galileo legend of popular imagination has little to recommend it, beyond sanctioning an unscientific attitude towards the Latin Middle Ages. Without doubt the seventeenth century can be credited with the birth of modern mechanics, and Isaac Newton can be credited with creating a new physical theory in the sense that his mathematical approach to nature successfully embraced the whole of physical phenomena, terrestrial as well as celestial. But this new mathematical theory of nature cannot be appreciated properly without understanding the two thousand years of scientific thought which preceded it. Nor can this new mathematical theory be evaluated intelligently without understanding the older, purely natural approach to nature which was peremptorily dismissed in the excitement.

We sketched briefly the Platonic theory of physical science as it prevailed in the Mediterranean world of the second and third centuries of the Christian era. Four centuries before Christ, Plato and his disciples sought to find the true explanation of natural events in the mathematical structure of eternal ideas existing apart from the world of matter. The Greek Fathers of the Church were not particularly interested in the mathematical developments of Euclid, Ptolemy and Archimedes, but they adapted Plato's cosmology to a Christian view of creation and the nature of the universe. With the fall of the Roman Empire practically the whole of Greek science, its mathematics, astronomy, mechanics and philosophy, was lost, and our European ancestors were left to work out a meagre existence and a negligible education, except for the truths of Christianity. The awakening of Western thought was due to the recovery of Greek science and philosophy in the twelfth and thirteenth centuries through translations from Greek and Arabic. In the thirteenth century there were two clearly distinct theories of physical science. The older Platonic view systematically subalternated natural science to mathematics, and mathematics to theology; in this view an explanation of natural events was ardently hoped for in mathematics, particularly in geometry. The other systematic view was the new Aristotelianism of Albertus Magnus, which defended the autonomy of natural

science, considering mathematics an introductory dis-
cipline and a useful tool in the scientific investigation of
natural causes. This purely physical theory of nature
was enhanced by the labours of Albertus Magnus and
by the profound clarity of St. Thomas Aquinas. This
approach to nature has continued to develop to the
present day, particularly in the biological and geological
sciences.

In the early fourteenth century Aristotelian physical
theory was to a certain extent corrected and developed
by John Buridan and his disciples at Paris. Some years
earlier two new approaches to nature had been inaugurated at Oxford. The first was a linguistic and grammatical approach to scientific problems, an approach
which conceived everything except individual substances
and qualities to be a mere grammatical term; this was
introduced through the theology of William of Ockham.
The other innovation was a new mathematical approach
which attempted to unite all physical motion, rectilinear
and rotational, locomotion and alteration, in a single
mathematical law of dynamics. This ideal was proposed
by Thomas Bradwardine, Fellow of Merton College; it
remained for his disciples at Oxford to establish the
problems and the terminology of the new "calculations
of motion". These problems and "calculations" laid
the foundation for the successful development of mathematical theory in the seventeenth century, when the ideal
of Bradwardine found temporary expression in the
universal mechanics of Isaac Newton.

Much has been lost by eliminating a purely physical
theory of nature from science. The biological sciences
have never been successfully incorporated into a mathematical theory; they cannot be swept into a mathematical
abstraction without serious loss. Hence many parts of
modern science are left without a unifying physical
theory. Moreover, modern physics is becoming an
increasingly complex attempt to account for the known

facts, and this "mathematical way" has need of theoretical complementarity in a physical theory of nature such as that suggested by Albertus Magnus. It is pleasant to think that the patron of those who cultivate the natural sciences can still speak to the modern world, and can still offer a guiding hand to the modern scientist.

On the whole, the great classics of Antiquity and the Middle Ages are to be preferred to modern summaries, although frequently a good modern paraphrase can present the ancient thought more vitally to non-specialists than the original. The works of Plato and Aristotle are readily available in a number of English versions; the basic works of Greek science and modern classical physics can be found in any good library, and often in modern translation. For medieval physical theory and detailed science, however, very little is available in any modern language. The works of Albertus Magnus and the philosophical writings of St. Thomas have been sadly neglected.

General Introductions

DUHEM, Pierre, *Le Système du monde, histoire des doctrines cosmologiques de Platon à Copernic,* Paris, 1913-17 (reprint 1954), 5 vols.; vols. 6-10, Paris, 1954-9. This pioneer work is still of fundamental importance as a general history of science, even where it must be corrected by more recent studies; eminently readable.

THORNDIKE, Lynn, *A History of Magic and Experimental Sciences,* 1923-58, 8 vols. Highly informative on astrological and related treatises from Roman times to the seventeenth century; a goldmine of MS. information.

SARTON, George, *Introduction to the History of Science,* Baltimore, 1927-48, 3 vols. (two parts to vols ii & iii). A biographical and bibliographical reference library for the centuries between Homer and A.D. 1400.

CLAGETT, Marshall, *The Science of Mechanics in the Middle Ages,* Madison, 1959. An excellent, detailed history of the mechanical sciences from Aristotle to Galileo.

CROMBIE, A. C., *Augustine to Galileo. The History of Science A.D. 400-1650,* London, 1952; 2nd ed. published as *Medieval and Early Modern Science,* New York, 1959, 2 vols. The second edition is a most readable and extensive presentation of all aspects of medieval and early modern science; profitable for the beginner.

DIJKSTERHUIS, E. J., *Die Mechanisierung des Weltbildes,** Springer-Verlag, Berlin-Heidelberg, 1956. An adroit summary of recent research in scientific thought from Plato to Newton, with frequent new interpretations of historical data.

Select Period Studies

SARTON, George, *A History of Science,* vol. i, *Ancient Science Through the Golden Age of Greece,* Cambridge, Mass., 1952; vol. ii, *Hellenistic Science and Culture in the Last Three Centuries B.C.,* Cambridge, Mass., 1959.

CLAGETT, Marshall, *Greek Science in Antiquity,* New York, 1955. On the whole a good introduction to Greek science, including the patristic period; somewhat weak, however, on philosophical thought and scientific theory.

HASKINS, Charles H., *Studies in the History of Mediaeval Science,* 2nd ed., Cambridge, Mass., 1927. Still the best discussion of twelfth-century science and translations.

CROMBIE, A. C., *Robert Grosseteste and the Origins of Experimental Science, 1100-1700,* Oxford, 1953. Bibliographically erudite and generally informative, but not systematically illuminating for most beginners.
* English translation in preparation.

GARDEIL, H. D., *Introduction to the Philosophy of St. Thomas Aquinas*, vol. ii, *Cosmology*, St. Louis, Mo., 1958. An excellent, brief summary of St. Thomas's general physical theory for beginners.

SMITH, Vincent E., *The General Science of Nature*, Milwaukee, 1958. An excellent modern presentation of St. Thomas's general physical theory.

DUHEM, Pierre, *Études sur Léonard de Vinci*, Paris, 1906-13 (reprint 1955), 3 vols. The classical pioneer work on fourteenth-century physics; eminently readable, but needs to be corrected by the works of A. Maier and others.

MAIER, Anneliese, *Studien zur Naturphilosophie der Spätscholastik*:

    I. *Die Vorläufer Galileis im 14. Jahrhundert*, Rome, 1949.

    II. *Zwei Grundprobleme der scholastischen Naturphilosophie*, 2nd ed., Rome, 1951.

    III. *An der Grenze von Scholastik und Naturwissenschaft*, 2nd ed., Rome, 1952.

    IV. *Metaphysische Hintergründe der spätscholastischen Naturphilosophie*, Rome, 1955.

    V. *Zwischen Philosophie und Mechanik*, Rome, 1958.

Absolutely indispensable for an understanding of physics and natural philosophy in the late Middle Ages; somewhat difficult reading for most students.

BURTT, E. A., *The Metaphysical Foundations of Modern Physical Science*, 2nd ed., London, 1932. The classical and unsurpassed study of the philosophical and religious basis of seventeenth-century scientific thought. (Paper back ed., New York, 1955.)

KOYRÉ, Alexandre, *From the Closed World to the Infinite Universe*, Baltimore, 1957. Excellent presentation of cosmological thought from fifteenth to seventeenth century with its metaphysical implications.

KOYRÉ, Alexandre, *Etudes galiléennes,* Paris, 1939, 3 vols. Best study of Galileo's mechanics and its immediate sources.

WHITEHEAD, A. N., *Science and the Modern World.* many editions. A well-known and important examination of seventeenth-century science in the light of modern relativity physics and contemporary philosophy.